AF825383

STRATEGY & TACTICS STAFF STUDY Nr. 2

Wargame Design

The History, Production, and Use of Conflict Simulation Games

by the Staff of Strategy & Tactics Magazine

SIMULATIONS PUBLICATIONS, INCORPORATED
New York

ISBN: 0-917852-01-X
Library of Congress Catalog Card Number: 77-081183

Copyright © 1977 by Simulations Publications, Incorporated,
New York, New York. All rights reserved.
Published 1977. First Edition.

Printed and Manufactured in the United States of America.

Note: Some of the material in this book appeared in a somewhat different form in issues of *Strategy & Tactics* Magazine and *MOVES* Magazine, publications of Simulations Publications, Inc. All of the typography in this book is newly set in Times Roman and Univers 55.

Wargame Design

Strategy & Tactics
Staff Study Nr. 2

CONTRIBUTORS
Richard H. Berg
James F. Dunnigan
David C. Isby
Stephen B. Patrick
Redmond A. Simonsen

EDITORIAL STAFF
Richard H. Berg
Robert J. Ryer
Redmond A. Simonsen

BOOK DESIGN
Redmond A. Simonsen

LAYOUT AND PRODUCTION
Manfred F. Milkuhn
Harry L. Park

Introduction

This is a book for the person who is more than casually interested in the hobby as well as the theory of designing conflict simulations, or wargames. The idea of designing one's own game is certainly not new. From the very beginning of the hobby, back in the early 1960's, there were at least a dozen variants on the old Avalon Hill standards, variants in which the fledgling designer attempted to improve on the original — usually with mixed success. There were also the dozens of "fanzines", many of which included the crudest of home-grown games: hand drawn hexes, counters and charts were par for the course and often provided a great deal of comic relief.

Yet the gamer was showing that he had more than a "playing" interest in wargaming. There was a conscious desire on the part of many to do their own game, to improve on what had been done. As the sophistication of the industry grew the desire to improve became the desire to create, and nowadays each year brings forth at least ten-to-fifteen privately designed and published commercial games.

These games vary considerably in their levels of excellence. Some amateur designers do their work so well that they can survive on their own as companies. Battleline Publications comes to mind as an outstanding example of this type of effort. However, most amateur-designed games are what are known as "vanity publications" — games published not to make money but simply for the designer to get his idea into print (or perhaps just his name!).

And the problem for most amateur designers is they just do not know how to go about actually doing what they want to do. The very first step — getting a map and reducing it to a game-board — will stymie all but the hardiest souls. Questions of how to design a CRT, what formulae are used to determine how many men equals one strength point, and why rough terrain is twice as difficult as clear may seem simple to the professional, but to the amateur they are so many Gordian Knots to untie.

This book attempts to answer those questions and then go several steps further. It first provides the gamer with a complete history of the hobby and the major personalities within it. This section is quite useful as it provides a background for the gamer and fledgling designer. Then there are the several sections which show the amateur how a game is designed. Sections cover the physical aspects of putting a game together: where to get materials, what to get, and how to use them; sources for research and where to find them; and even a step-by-step run-through of a game being designed, so that the amateur can see the philosophy and reasoning behind the moves that were made. There are further sections which deal with various problems that will arise during the course of design, and how they were — and can be — handled. There are also practical sections on finances (the business end), usage of military symbols and playtesting. Finally, there is a listing of wargames available (and not available) and the major companies that supply them.

In all, here is a book that will provide a goldmine of information to any gamer more than peripherally interested in the hobby. It provides insight and practical knowledge, and with a little creativity and research, you might even find yourself an "amateur" designer yourself some day.

SECTION I:

The History of Wargaming

by Stephen B. Patrick

THE STORY OF WARGAMES

Wargames are probably only slightly younger than organized warfare. Archeologists have uncovered a group of Sumerian warriors, marching in miniature phalanx, and a similar group of Egyptian miniature soldiers has been found arrayed in full panoply. It is probable that the sons of the nobility—the military leaders in ancient times—received their initial training with such miniatures. At some point, the play associated with these miniatures became stylized. From this stylized pastime sprang the true games that we play today, often without recognizing their military origins: Chess, Go, and kindred games. As a result of their development, wargames developed along dual lines, as a means for both military training and amusement.

Military Wargames

Military wargames (i.e., those played by professional soldiers or policy makers while on the job) can be distinguished from *civilian* wargames (i.e., those published by such people as Avalon Hill and Simulations Publications) in several ways. The most important distinction is purpose. While the civilian wargame is designed for entertainment and for historical interpretation, the military wargame is intended for training and for predicting possible real futures. Sometimes the tie to the future is very explicit, calling for soldiers to leave the gameroom to try to repeat the game results in real battle. In other instances, the purpose for playing a particular game is more general. But in any case, the purpose of the professional military game is to prepare for future crises.

BACKGROUND

The roots of such wargames lie far in the past. Chess, Go, and even Checkers were probably once thought of as suitable training for war. But if Chess had ever possessed any value as a simulation, that value had long since disappeared by the 18th Century in Europe. By that time, there had been too many developments since the days when battle was simply a collection of individual combats. However, Europeans of the 17th and 18th Centuries still were possessed of a desire to understand their environment scientifically and, for that reason, several thoughtful people decided that the ancient game of Chess should be updated to make it useful once more for understanding war. Indeed, as early as 1644, one Christopher Weikhmann invented what he called "The King's Game," a 31-piece development of Chess. There were many other games with assorted pieces and rules that were referred to collectively as "Military Chess" or "War Chess" games.

The first game to go far beyond its chess origins, however, was a game invented in 1780 by Helwig, Master of the Pages for the Duke of Brunswick. Helwig's game was played on a board of 1666 squares. The squares were colored in various ways to represent different types of terrain and also contained numbers which represented smaller terrain features. Each army defended a fortress, and the game was won by capturing the enemy fortress. Helwig added something new by providing that his pawns represent units of men instead of individuals. Each side was given 120 fighting units, including infantry, cavalry, and artillery as well as some pontoons and 200 entrenchment counters. The rules provided for such specialist pieces as pontoniers. The movement allowances were 8 spaces for infantry, 12 spaces for heavy cavalry, and 16 spaces for light cavalry. Helwig's game became quite well known and was introduced throughout Western Europe.

A more complex type of game that operated on the same basic principles was developed in 1795 by Georg Vinturinus, a military writer in Schleswig. In 1798, Vinturinus adapted his game to a new board—one that was based on an actual area, the border between France and Belgium, one of the most fought over areas in Europe since the 17th Century. The board had 3600 squares. In addition to fighting units, there were numerous military stores and devices to keep track of, such as siege equipment, fortifications, bridges, bread. In accordance with the

18th-Century idea of the importance of lines of communication, armies were required to maintain a line to their bases.

None of these early games had any particular value as aids to military training. They were like the historical board wargame of present—useful to visualize military concepts, excellent as entertainment for those with the time and the patience to understand their rules, but not much of a preparation for actual military experience.

A Prussian father and son were the inventors of a series of games that won the respect of some Prussian officials and started the wargame to its place as a mainstay of military training in the 19th Century. The father, Herr von Reisswitz, was a civilian who first developed his game in the dark years of Napoleonic domination in Germany after the Prussian defeat at Jena. He moved toward greater realism by dispensing with a grid pattern for movement and by using a sand table to represent terrain. Furthermore, he adopted a particular scale (1:2373) and attempted to bring everything into harmony with that scale, including the sizes of the unit counters—blocks of wood with military symbols pasted on. In 1811, Reisswitz's game was shown to two young Prussian princes who arranged an interview with the King. For that occasion, Reisswitz made a deluxe version with plaster terrain and porcelain units. This game became a constant plaything at the Prussian court, and from there it spread to the Russian court. But it was not used by soldiers until Reisswitz's son took an interest in it.

THE RISE OF MILITARY WARGAMES

The younger Reisswitz was a first lieutenant in the Prussian Army in 1824, when he began to experiment with his father's game. First he transferred the game to realistic military maps with a scale of 1:8000. He published a set of rules in 1824 and supplements in 1825 and 1828. Meanwhile, he had impressed some of the highest officers in the army with the game's usefulness. In 1824, the game was played for General von Muffling who was then Chief of Staff. Muffling was skeptical, but he consented to witness a demonstration and was thereby converted. "It is not a game at all," he exclaimed. "It's a training for war! I shall recommend it most emphatically to the whole army." And he did. Each regiment was furnished with a set and officially urged to practice on it.

The game developed by the younger Reisswitz was more realistic in several ways than earlier games. The terrain was represented on terrain maps drawn exactly as real military maps would be drawn. The players were not allowed to see the actual situation on the map. Instead, the services of an umpire were required to keep track of the situation on the map and to tell the players what they could see or what their patrols could discover. A time scale was accurately maintained with each move being equal to two minutes of real time. The tables furnished with the game provided for movements in this time period. Each play of the game could be different because it was made up by the umpire or someone else outside the game. The opposing players were given the same information on the general situation but were given different tasks to accomplish and different forces to use. The special information also included limited information on enemy forces and possible missions. For each move, written orders were given to the umpire who harmonized them and reported back the discoveries of troops on the move. However, the individual players were represented on the map and discoveries were not reported to them until a messenger had had time to reach them. Each side could be represented by a team instead of a single player, and then the players on the same side could not communicate with one another directly unless they were close together on the map. When the two sides clashed, the umpire determined losses by throwing dice and consulting an odds table.

Reisswitz's game was enthusiastically recommended by the Prussian Chief of Staff, but that did not make it popular among very many of the officers, nor did it make its inventor popular. He was transferred to the boondocks by jealous senior officers and harped at for inaccuracies in his game by most of those who played it. Consequently, he became despondent and committed suicide in 1827.

But the game had its fans in the Prussian army. One of them, as early as 1828, was Lieutenant Helmuth von Moltke, future Chief of Staff and victor in two wars. Enthusiastic players combined to form a wargames club, the *Kriegspieler Verein,* and to publish the first wargaming periodical. All of the wargame's devoted players changed the game as they played it. Several Prussian officers published their own sets of rules in attempts to make the game more realistic. The general tendency was to loosen the rigidity of the rules and to increase the responsibility of the umpire. Many changes were made, particularly in the computation of losses which was felt to be quite rigid and arbitrary in Reisswitz's original game. The normal method came to be a standard and multiplier system. The casualties sustained in some simple situation were determined and became the "standard." Then the percentage change resulting from different circumstances was used to vary the standard. There would be a different multiplier for each circumstance that differed from the standard circumstance. This system made the game more realistic, but it also made the game considerably more tedious as many multiplications had to be made to figure losses. Still, the game was considered too detailed *and* too unrealistic by most officers. It became more popular, however, when Moltke pushed it from the top after becoming Chief of Staff in 1837.

FREE KRIEGSPIEL OPPOSES RIGID KRIEGSPIEL

In 1876, Colonel von Verdy du Vernois, leading military writer and instructor, expressed the dissatisfaction of many officers by calling for the simplification of wargames. Basically, Verdy advised that most of the rules and the dice be eliminated but that the basic idea of the game be retained under the direction of an umpire experienced in actual warfare. Of course, by this time, Germany had been victorious in the Seven Weeks War and the Franco-Prussian War and had many experienced officers. In effect, then, the umpire would make up the rules and apply them as he went along, and the players would have the freedom to attempt things that might or might not be allowed by the umpire. Aside from the new importance of the umpire, Verdy's game was played like the other kind. The situation was variable, the players were separated and given only the information they could legitimately possess. A few standards were adopted as a basis for movement. For example, a battalion in line formation occupied a front of 200 yards, and fresh infantry could march on good roads at a speed of 117 yards per minute. But it was up to the umpire to make any changes in these standards due to the exhaustion of the men represented or the badness of the weather or roads.

The suggestion of Verdy du Vernois led to a division among wargamers, creating a rift between those who favored the altered game—called free Kriegspiel—and those who favored the traditional game—called rigid Kriegspiel. Both versions could be applied to any size game, but the complexity of the calculations necessary for the rigid game usually limited it to the representation of companies and battalions. In both its forms, it became a mainstay of training not only in the German army but in most of the world's armies.

The Germans impressed the whole military world in 1870 with their victory over the French, previously considered the best soldiers in the world. As a sign of German ascendancy, many armies, including that of the U.S., abandoned their French-style headgear—the *kepi* worn by Union soldiers in the American Civil War—and adopted the Prussian-style spiked helmet. More importantly, they investigated and adopted German staff procedures, German organization, and German training methods. Along with everything else came the German *Kriegspiel*. Apparently foreign soldiers gave wargaming more credit than it deserved for the German victories. But with the availability of Free Kriegspiel during the period of tense peace between 1871 and 1914, all armies including the German made wargaming a more important part of their training program.

In England, for example, wargames were introduced with the publication by the War Office of a game called *Aldershot,* a rigid Reisswitz-type game. In 1883, the commander in chief, the Duke of Cambridge, issued official orders recommending wargaming, and in 1896 a set of rules was published in a service journal. In England, wargames were often played on military maps of England and arranged so that the climax of the encounter would take place in an area near the post where the game was being played.

The Italian army favored rigid Kriegspiel and made its play an important part of the course at the War College. Forty evenings were spent learning the game and various versions were played over a two year period. It was intended mainly to instruct in logistics and other staff work. In the French army—the defeated army of 1870—there was some interest in wargaming, but the French were much slower to adopt this device than other soldiers, and it was not until 1900 that it became widespread.

Both the Russians and Japanese armies used wargaming, but it was accepted less enthusiastically by the Russians who felt that it was not worth the bother. When the Japanese defeated the Russians in the Russo-Japanese War of 1904, the Japanese gave much of the credit for their victory to their play of wargames, while the Russians turned to them with a new interest.

Rigid Kriegspiel was introduced into the American army in 1867 through the translation of a German work. It spread slowly but inspired two American officers who independently devised their own games by the 1880's. Major William R. Livermore published *The American Kriegspiel, A Game for Practicing the Art of War Upon a Topographical Map* in 1882. Livermore's major contribution was to recommend a series of devices to cut down on the paper work involved in playing a rigid Kriegspiel. The unit counters were marked on four sides with stripes and dots to represent strength. Other blocks, called "counters," were provided to be placed in front of troop units to show fatigue. Other blocks called "checks" were used to show disorganization following a battle. Assorted pointers, called "indices," were to be placed on the map to indicate fire and movement. Computations of the effects of fire were supposedly simplified by a "Firing Board"—a pegboard used to keep track of time and to help in figuring out casualties by the movement of pegs in the holes.

At about that same time, Lieutenant Charles A. L. Totten was working on his game, published as *Strategos: A Series of American Games of War* in 1880. Totten's game is distinguished by his attempt to appeal to the amateur as well as the professional. He provided a highly complex game similar to Livermore's, but he also provided a basic game called "The Battle Game" which used a square grid map to simplify movement.

Because Livermore's and Totten's games were so difficult and inconvenient, they met the same resistance in the American army that other rigid Kriegspiel rules had encountered overseas. Livermore attempted to disarm some criticism by recommending that the umpire disregard the charts and estimate movement and fire effect whenever possible. Free Kriegspiel, based on the work of Verdy du Vernois, was also introduced into the United States, but it was also criticized. Several officers argued that free Kriegspiel replaced arbitrary written rules with even more arbitrary unwritten rules, and that few men had the authority to be an umpire under such a system. By the end of the century, there seemed to be a tendency for the two systems to coalesce into one, becoming semi-rigid (or semi-free) Kriegspiel. Officers who advocated free Kriegspiel were found on occasion to consult charts and rules, while Livermore, identified with the rigid game, was reported to disregard his own tables and charts as often as he consulted them.

After 1900, discussion and invention of wargames continued in the armies of the world. Generally, since there were few new experiences (and since such experiences as the Russo-Japanese War were generally misunderstood), there were few changes in wargaming. Books published by German Captain Frederick Immanuel in 1907 and by American Major Farrand Sayre in 1908 were among the best contributions to the discussion. Immanuel was in the free Kriegspiel tradition. He vigorously preached the doctrine of flexibility and all power to the umpire. Nevertheless, with the problems of the beginning umpire in mind, he did give some recommended distances for movement, based on a move interval of 2½ minutes. Infantry, on roads, would move 1,000 meters in 12 minutes. Cavalry and Field Artillery would walk 1,000 meters in 10 minutes but would gallop the same distance in only 2½ minutes. He also provided a table for deployment. An infantry battalion would deploy in 5 minutes, a regiment in 15, and a brigade in 30. Immanuel recommended several ways to speed up the game and make it more interesting and instructive. Unimportant skirmishes could be ignored, preliminary maneuvering could be eliminated, or extra information could be given out in order to bring the opposing forces together faster.

Sayre's game was more in the rigid Kriegspiel tradition, but he too advocated flexibility. Thus the major difference between his book and Immanuel's was that Sayre supplied a table for calculating fire results as well as a table of march distances. Under good conditions and over long distances, infantry moved at 81 yards per minute and to 110 yards per minute. Several factors could lessen the speed. Sayre made a good provision for easing the figuring of casualties in fire fights. By using multiples of approximate logarithms, the painful process of multiple multiplications could be changed to the faster process of addition. For rifle fire, the table took account of the following variables: in regard to the troops firing, effectiveness could be changed by two possible rates of fire, three positions of firing, three degrees of skill, freshness or fatigue, three levels of morale, and according to the amount of fire being received; and in regard to the target troops, the variables were four angles of front, five amounts of motion, ten formations, and seven positions. These variables are numerous, but they are fewer than those provided by Livermore or Totten and more easily if less accurately handled.

STRATEGIC WARGAMES

The late 19th Century also saw the growth of strategic wargaming among all the world's armies. Generals were having to plan for more and more men armed with more and more sophisticated weapons and traveling by railroad. The French disaster in 1870 showed the necessity of a smooth-working plan for mobilization and movement, and the situation grew in complexity every year after 1870. The generals would have preferred plans that had proved their usefulness, but this could never be done in the real world except through war, and by then it would be too late. So the generals turned to games to prove their designs. Since the matter was so complex, these games were always based on the free Kriegspiel approach rather than the rigid.

The Germans, as usual, were in the lead. One of the most avid wargamers in the world at the time was the Chief of the German General Staff from 1892 to 1906, Alfred Graf von Schlieffen. Schlieffen relied on the results of extensive wargaming to develop and revise his plan—essentially the same plan with which the Germans entered World War I. However, the games used to test the Schlieffen plan were no better than the ideas which went into them and were not able to point out its faults. The whole design assumed that the Belgians would not fight seriously, that the British would not land in force, and that the French would be incapable of transferring men from the right flank to the left. The games then also showed these things. Nevertheless, in the actual campaign in 1914, all of these unplanned-for things happened and the opening campaign ended in a stalemate instead of the absolute decision that the German generals had hoped for.

The Russian wargames were never as systematic as the German, but they did get a potentially valuable piece of information by way of a game. In April 1914, they played a wargame on their invasion of East Prussia, an action they had promised the French to perform soon after the declaration of war. In the game, the Russian armies were commanded by the two men who would command them in case of war, Rennenkampf and Samsonov. The game showed that the two Russian armies were separated because of the terrain and were subject to defeat in detail. This was a valuable lesson, but it was not applied. Later that year, when the actual campaign was attempted, the original plans were used, the two armies were separated, and they were defeated in detail at the battles of Tannenberg and the Masurian Lakes.

In contrast, the British were able to learn from a game they played in 1905. They played a German invasion of Belgium opposed by the Belgians and by a British Expeditionary Force. The game demonstrated an important lack in British preparations for such a war: there would not be sufficient transport available on the outbreak of the war to move the British army fast enough to help the Belgians. The game led to the increase of transportation preparations and also led the British army to begin unofficial staff talks with the French army, a trend that helped to bring Britain into the war and to make its entry effective in 1914.

After World War I, strategic wargames continued to be relied on by the general staffs. The German army added something important to the wargaming tradition in 1929. This was during the period of the 100,000 man German army, and Germany had to depend more on moral force than physical force for defense. So, following a suggestion from Manstein, the game began with a period of tension between Germany and Poland—Poland was the aggressor—and included the reactions of Britain, France, and the League of Nations, as well as military operations, first against Polish irregular forces and then against the invading Polish army.

Another game that had no issue in reality was played in 1938. As part of an effort to prevent Hitler from starting a war between Germany and Czechoslovakia, General Beck, the army Chief of Staff, conducted a wargame. The game demonstrated that the German army could beat Czechoslovakia only at prohibitive cost and with disasterous long-term results. Thanks to the Munich agreement, there was never any reality to compare with the results of the game.

Most German wargames of the Hitler period were strictly military and generally proved trustworthy in the narrow military sphere. All of the German army's better prepared attacks were thoroughly wargamed before the fact. During the so-called "phony war," the Germans carried out a particularly important game that showed that the march through the Ardennes could be made fast enough to surprise the French. In the event, the march was finished in even less time than had been planned. Later in the year, games on Operation Sea Lion showed the many difficulties obstructing the proposed invasion of Great Britain and contributed to the reluctance of the Germans to actually carry it out. Then Operation Barbarossa, the plan for the invasion of Russia, was the subject of extensive wargaming which contributed greatly to the speed and success of the first few weeks' battles.

One of the best known wargames of World War II was one that the Japanese used in planning their proposed Midway operation. The game's plan took into consideration all of the known variables that might have bearing on the operation's outcome. What resulted was a matrix of probable U.S. forces available versus known Japanese forces and of probable U.S. tactics in the face of planned Japanese tactics. When play ended, the Japanese had lost the game. Since they had surprise and numerical superiority in their favor, this outcome was considered unacceptable. The rules of the game were adjusted (in the tradition, perhaps, of all wargamers) to produce the desired Japanese victory. Having vindicated themselves on the game map, the Japanese then proceeded to substantiate the verdict of the initial game design by suffering their first major defeat of the war.

One of the most remarkable German wargames began on November 2, 1944. The staff of the 5th Panzer Army defending Germany against the Western Allies anticipated an American attack on their front by wargaming it. Part way through the game, the Americans actually launched the assault. Army group commander Model ordered that the game be continued. Information from the front lines was fed into the game and the defensive orders for the game were transmitted to the front for application in the real battle.

WAR GAMES SINCE WORLD WAR II

Since 1945, every kind of traditional wargame has been played by the armies, navies, and civilian military planners of the world. Furthermore, there has been a revolution in wargaming that can only be sketched here. As in previous periods, there has been a rapid technological revolution in military hardware since World War II. Furthermore, one large class of weapons—nuclear missile weapons—has never been tested in any trustworthy fashion. But even with more conventional weapons, the wars that have been fought have given very lopsided experience with the use and combination of weapons. So military men have continued to play games in an attempt to ponder the imponderable. Naturally, the wargaming world itself has participated in the technological revolution.

Because of the new technology, there has been a definite reversal of the trend toward free Kriegspiel, which was predominant from 1900 to 1945. High-level wargamers have large funds at their disposal and use a wide variety of calculators, computers, and sophisticated electronic communications equipment in their business.

One of the more traditional games is that played in the Marine Corps, called the Landing Force Game. It can be played with any number of situations or scenarios. It is a rigid game with a set of rules that fills two thick volumes. Although it uses an electronic random number generator instead of a pair of dice, it is still very close to the late-19th Century rigid Kriegspiel, and it takes about four months of real time to play 30 hours of game time. Other contemporary games are only a short step away from the traditional since men still make the decisions, although computers are used to calculate movement and the effects of combat. These games proceed much more rapidly than the rigid noncomputer games, but they operate on the same principles.

However, some games now are played entirely by computer. The all-computer game is extremely rigid since rules for every possibility have to be fed into the computer before a game can begin. The advantage of the all-computer game is that it can be played many times in succession with minor variations. Thus the designers of these games, or *models,* can try to use them to discriminate between competing strategies or weapons systems.

The United States has now become the center of wargaming, and the U.S. defense establishment has witnessed the playing of games on every possible defense subject with its varied arsenal of gaming techniques. Foreign policy crises have been gamed in attempts to guess how the armed forces might become involved. More straightforward military actions are also gamed, but these too might have non-military complications. One of the most active areas of gaming is that of nuclear exchanges. Generally, games are used for research in areas that cannot be easily penetrated by other methods and for the training of people in crisis thinking, both political and military. The subjects of such games range from the immediate future to the possible worlds of 1984.

The U.S. Army's extensive use of wargames runs the gamut from sand table exercises conducted at platoon and company level, to Command Post Exercises (CPX) usually conducted at battalion level and above, to actual on-the-ground Field Training Exercises (FTX) usually held up to Corps level. The Army FM 105-5, Maneuver Control, provides a wealth of material for a planned exercise in terms of controlling the realism. It contains such data as rates of advance based on the type of attacking force and the type of defender. Of course, the United States is not alone in running such wargames. They not only permit rehearsal of planned operations, but they also provide varying degrees of training to suit such situations as limited maneuver area or other cost restrictions which may completely rule out using actual troops.

There is a vigorous debate in and out of the defense structure about whether all this wargaming is worth the cost and effort put into it. The more traditional wargame has probably proven its usefulness in its limited sphere, but there may be some danger in the extension of the game idea to matters of general foreign and military policy. Frequently, the only reason that planners continue to play games is not that they trust games to give them legitimate answers to their problems, but that they cannot think of any way to get such answers. Maybe, they hope, next year's games will be better.

However, in 1976, under contract from the U.S. Army, SPI designed a game especially for use in training lower echelon officers and upper enlisted men in company level tactics. The game, *Firefight,* used information from various sources and was designed to the specifications of the Army. It is, though, available to the public.

Non-Military Wargames

As noted previously, the point at which wargames diverged into those intended for training and those for amusement is lost in the mists of time. Chess and Go are the two most visible and popular products of this division. In addition, wargames with miniature soldiers have been a traditional pastime for hundreds of years.

MINIATURES

The senior partner in the non-abstract wargame field has always been miniatures. Young boys in well-to-do families were expected to play with soldiers because, in former times, younger sons had no prospect of inheriting the family wealth and estates, which went to the eldest son. Younger brothers had to carve out their own careers, and it was a normal course for them to enter the military. Such a career is typified in that of the third son of the First Earl of Mornington. Through skill as well as his father's influence and wealth, this young man became a battalion commander by the time he was twenty-three. His eldest brother succeeded to the family title, and he was forced to content himself with a military career. This third son, Arthur Wellesly, had the good fortune to select a career for which he was eminently suited. He eventually became Britain's greatest commander since Marlborough—the Duke of Wellington.

It is this same tradition that gave us Winston Churchill who, as a boy growing up in the home of his uncle, the Duke of Marlborough, played with his case of toy soldiers. Churchill could not aspire to a family title, so he sought his fame elsewhere.

Despite this tradition, there is little evidence to indicate any serious, organized rules for playing with toy soldiers. Local ground rules were the only rules, as in the case of Churchill insisting that his brother use the African troops without artillery while Winston helped himself to the regulars and all the artillery. As a serious hobby, wargaming with miniatures can be traced to its legitimate beginnings with H. G. Wells' *Little Wars.* The man who gave us *War of the Worlds* published, in 1912–1913, several articles which gave the first formal rules for wargaming to be read and used on any sort of wide scale. From these rules have developed the wide range of rules for miniatures that now exist, covering everything from ancient warfare to modern times.

Miniaturists have generally fallen into two groups. First, there are those for whom play is secondary and authenticity primary. They may have small armies, but each soldier is painted in the precise manner of the uniform of that particular unit. As a general rule, this group of hobbyists will use the larger size pieces—54mm or 1:32 scale. For those miniaturists who are more concerned with play, smaller pieces are more acceptable because they are less expensive to purchase in quantity.

Although miniatures depict individuals, each figure by convention is normally used to represent a number of men. For instance, one miniature soldier is commonly used to represent twenty men. This ratio results in a battalion that is more manageable to deploy. The use of miniatures stimulates an interest in linear tactics and the periods in which the tactic of maneuver was at its height. Consequently, most miniaturists are interested in the period from about 1740 to 1815, although there are sizable contingents who are interested in other periods as well. The Battle of Waterloo is, without a doubt, the most re-fought battle of them all.

Another attraction for miniaturists is the myriad petty German states that sent mercenaries around Europe in the latter part of the Eighteenth Century. This proliferation of men in arms allows hobbyists to field their own favorite regiment, creating great—and attractive—diversification in decoration and painting.

Perhaps even more than their board-playing counterparts, miniaturists tinker with rules. There are no rules that are uniformly agreed to as the best rules for a given era or even a given battle, and there are as many "ultimate" rules as there are points of view on what makes a good set of rules.

Naval miniatures focus more around rules created by Fletcher Pratt, although these rules are also something less than sacrosanct. In contrast to their land warfare counterparts, naval miniaturists tend to favor more modern periods, choosing the dreadnought era over that of the sailing ships as a rule. Thus, for naval people, it is Jutland rather than Trafalgar that is refought.

Only aerial warfare is really slighted among miniaturists. Although there are rules for adding aerial elements in land warfare games of the modern era, there seems to be little interest in aerial miniatures—a phenomenon which is most likely due to the very nature of what is being depicted.

BOARDGAMES

The growth of published games, such as *Monopoly,* did not see a comparative effort to create wargames for general consumption. Parker Brothers did publish *Camelot,* but without any effect as to opening up the wargame field. There seemed to be an idea, lasting into the post-World War II period, that an intelligent wargame was strictly the prerogative of military minds. Despite the fact that World War II had held the interest of everyone down to children, no serious effort was made to exploit this interest by putting out a serious wargame.

The cost of miniatures, in conjunction with their limited availability and the complexity of even the simplest rules for their use in games, kept the hobby of wargaming small. The transition of wargaming from a limited audience to a mass hobby required a major breakthrough.

Modern wargaming on boards, as a hobby, can be traced to one man and one game. In 1953, Charles S. Roberts, a young man in his early twenties, combined an interest in the military and in history to produce a game, which he designed in his spare time, called *Tactics.* It dealt with two hypothetical countries, each of which had an army of the post-World War II type. The game was printed commercially, although not done on oil cloth as legend has told it. It was distributed through the Stackpole Company of Pennsylvania, a company better known as a publisher of books on military science. This was only a part-time pursuit for Roberts at this point, as he was still working in the advertising and marketing field for a living. On the other hand, Roberts had put considerable time and effort into the design of the game, as well as its sale and distribution. He had concluded that there would be a market for such games. As a result, he had an idea of what lay ahead. Between 1953 and 1958, he sold about

2,000 copies of *Tactics* and, in his own words, "came within about $30.00 of breaking even."

Having produced and marketed this first game with some success, Roberts concluded that there was a market for a broad range of adult games. He expected to appeal to one segment of this range with *Tactics*. He also thought that other games could be sold which could be financially successful without appealing to a broad audience in the manner of *Monopoly*.

By 1958, Roberts was convinced that a living could be earned in the game design and sales field. At the same time, he found himself knee-deep in games at his home, where he had stored them. His wife objected most strongly and that, together with the success, caused him to consider designing and selling games on a more formal basis. In the spring of 1958, the Avalon Hill Company had been formed with Charles S. Roberts at its head.

AVALON HILL: YEARS OF GROWTH 1958–1962

The Avalon Hill Company was *not* founded for the primary purpose of producing wargames. This point is often ignored by those in the hobby who have come to look to Avalon Hill as a source of games. Its true purpose was, and remains, to produce the broad spectrum of adult games for which Roberts felt there would be a market. Roberts felt that the big game publishers, Parker Brothers, Milton Bradley and the like, had ignored the adult game field and he was determined to take advantage of their indifference. From the outset, Roberts made little effort to enter the popular game field of Parker Brothers and the rest. He priced his games at $5.00 when other companies were selling theirs for $2.00. This brought resistance from both wholesalers and retailers who thought that such a price would be too high. Roberts had the opinion that if people were interested in a game, "they would pay $5.00 for it; and if they weren't interested, they wouldn't pay fifty cents for it." This set the basic selling strategy for Avalon Hill and events proved him correct as sales began to grow rapidly. In order to sell games at $5.00 a copy, Roberts was forced to develop innovative marketing techniques, including a very heavy promotion program. His initial distribution was in Baltimore, but Avalon Hill soon began to sell nationwide as his idea proved successful.

Although the first two games, *Gettysburg* and *Tactics II,* were wargames, the next game, *Verdict,* was not. In fact, during the Roberts era, Avalon Hill published (in addition to *Gettysburg, Tactics II,* and *Verdict*) *U-Boat, Management, Chancellorsville, D-Day, Nieu Chess, Verdict II, Air Empire, Le Mans, Civil War, Baseball Strategy, Football Strategy, Waterloo, Bismarck,* and *Stalingrad,* plus *Word Power*. Of these eighteen titles, nine are non-battle titles representing such diverse fields as law, commerce, and sports. All of this reflects Roberts' desire to have a broad spectrum of games. Though one could not say a hobby had been born, it became clear that the most constant result could be obtained from the sale of battle titles.

The rationale behind a game such as *Gettysburg* is rather simple. *Tactics* was already out, under Roberts' own name, and the Civil War Centennial was coming. Roberts wanted to take advantage of the rise in interest in the Civil War. *Gettysburg* had a square grid system, just as *Tactics* had, though the square was soon abandoned for the hexagonal system used in other Avalon Hill games. It is interesting to note that an attempt to reissue *Gettysburg* with a hexagonal grid system was unsuccessful, and the version now sold is the square grid. *Tactics II,* which followed, was a reworking of *Tactics*. It also provided the advantage of a fully designed and tested game ready for production.

In the early fifties, when Roberts was selling *Tactics* through Stackpole, he was contacted by a representative of RAND corporation. After much comic-opera super secrecy, during the course of which things reached the point that the RAND man advised Roberts that they wanted to talk to him but couldn't tell him why because he, Roberts, was not cleared by security, Roberts learned that RAND wanted to know where he got his combat results table in *Tactics*. This is the same CRT used in all of the early Avalon Hill games. Further, it bore a resemblance to the more complex one that RAND was using to wargame World War III and other horrors. The humorous aspect, as Roberts notes, is that RAND probably spent thousands of dollars to come up with the CRT he developed in a simpler form in fifteen minutes. It was no more complex than the basic doctrine that a defender can hold off a force three times its size.

This episode, in turn, triggered Roberts' interest in RAND. One day he saw a picture of RAND's wargame room and noticed that they played on a hexagonal grid. Recognizing a good thing and having had troubles with the square grid, Roberts took it over.

Not long after Avalon Hill got under way, Roberts found that it was too much of a strain to handle on a part-time basis. He quit his advertising job and devoted his full efforts to making Avalon Hill grow. Before long, he turned to his former place of employment and recruited Bernard C. Schramm to fill a critical need at Avalon Hill. Schramm was experienced in the printing field, and his knowledge enabled Avalon Hill to avoid the trial and error method of obtaining the printing results they wanted. Schramm had virtually nothing to do with game design and devoted his attention to management and production.

Despite the addition of Schramm, Roberts found that he could not manage the company and still devote full time to game design. Again he turned to the advertising firm where he had worked and this time persuaded Thomas N. Shaw to work for Avalon Hill as a game designer. Shaw had previously designed two games which had been privately printed in the Baltimore area. Roberts saw them and liked them. The two Shaw-designed games were released by Avalon Hill as *Football* and *Baseball Strategy*.

Shaw arrived as Roberts was in the midst of expanding the line. Until his appearance, there were only six games in the line. The toy fair of Spring 1961 was Roberts' target date for doubling the size of the line. Shaw's first game was *Air Empire*, a game he designed based on Roberts' earlier *Management*. In addition, Shaw worked on *Verdict II*, helping to develop some of the cases used in the game.

During this period, Roberts took the lead in determining what games would be produced. No wargame produced by Avalon Hill during this period lacked his influence. Therefore, many of the credits for game designs must be reconsidered. *Gettysburg* and *Tactics II* were Roberts' design from start to finish. *Dispatcher* was likewise wholly Roberts' design. *Verdict* was designed by two attorneys and therefore represented the first outside design. *Management, Chancellorsville, D-Day,* and *Nieu Chess* were all designed by Roberts. *Verdict II* was, again, designed by the attorneys, with more direct influence by Avalon Hill in the form of cases being worked up by Shaw and the degree of difficulty being controlled by Roberts. *Verdict II* was a simplified version of *Verdict*, which it replaced. *Air Empire* was designed by both Roberts and Shaw. Roberts conceived the idea and influenced Shaw's work in putting the mechanics of the game together. *Le Mans* was another outside design, but *Civil War* was Roberts'. This was one of the few ventures Avalon Hill made into the popular game field. It was priced lower than the rest of the line and was to compete directly with Parker Brothers and the rest. Though not up to the standards of the regular wargames, it was a success and made a profit. *Waterloo* was the first wargame in which Roberts did not have a major hand. It was designed by Shaw and one of his principal assistants, Lindsley Schutz. On *Bismarck*, there is some difficulty in establishing who deserves the primary credit. Roberts says Shaw and Schutz did the majority of the work; Shaw says that Roberts sketched out the design and they just finished it off. The answer to the question of who designed *Stalingrad* turns on who defines the word "design." The idea was one that Roberts had had for a long time. The mechanics, by then, were "standard" Avalon Hill rules. Roberts had previously roughed out the game on several different occasions, and he turned it over to Shaw for final development. It was Lindsley Schutz who put all of the pieces together by getting the background information. In terms of work done, the game is Schutz's, but in terms of original idea and having the final decision on which way the game went, the game is really Roberts'.

Physically, Avalon Hill was not self-contained. Although the design was done in the "plant," the final art-work, printing, and other elements necessary for production were done outside. The two major outside companies were James Smith Box Company, which did the boxes and parts, and Monarch Services, which did the rest of the printing. Monarch now controls Avalon Hill as a result of the financial difficulties later encountered by Avalon Hill.

The playtesting of the games, particularly at the outset, suffered from too few playtesters. As a result, rules additions and corrections were a norm. In part, the problem was solved by making use of outside testers. These people were drawn from the audience which had built up rapidly around the games. They wrote letters, often incisively, and were not afraid to express their opinions. Avalon Hill would duplicate a quantity of "dummies" of the games and send them, with commentary sheets, to the more lucid of these correspondents, who were asked to point out any weaknesses they might find. As Roberts notes, this served two purposes. First, it cut down on the mail these people sent in about game mistakes since they had a chance to find the mistakes themselves before the games were printed. The second and more important purpose was that it gave Avalon Hill a handy, semi-controlled body of people who could "de-bug" games. Among the playtesters of this period was Louis Zocchi, who later went on to design such games as *Battle of Britain* and *Luftwaffe*.

Avalon Hill also explored the non-commercial wargame field. Roberts particularly tried to expand the educational value of the games, especially the wargames. He designed a game called *Gametrain* which he tried to sell to the U.S. Army Infantry School at Fort Benning, Georgia. The game was to be used for training an infantry platoon and was on a board measuring in excess of twenty feet to a side! When *Gametrain* was demonstrated to the Infantry School, it was accepted by the military men but rejected by the civilians who had the final say. *Gametrain* never reached fruition.

In the non-wargame field, Avalon Hill worked with the American Management Association to develop management training techniques, an ideal field in light of the simulation aspect of the games. Roberts has remarked, somewhat ruefully, that he would have been a lot richer had he not spent so much time in these ventures. Yet, despite these side efforts, by 1962 Avalon Hill was doing a million dollars per year in sales.

There were other projects. When a new buyer sent in the registration card found in each game box, he would receive a flyer asking whether he would be interested in a game costing $25.00 a copy. This game was only a general idea of Roberts. It was to be called *Conflict* and involve three hypothetical countries. Although *Blitzkrieg* was not an offshoot of this, the idea is not unlike a super-*Blitzkrieg*. *Conflict* never got beyond this stage because the response was never great enough to warrant full development.

In any event, at the end of 1962, Avalon Hill had virtually cornered the market on adult games.

AH: DISASTER AND REBIRTH 1963–1964

In 1962, Avalon Hill experienced the first signs of potential problems. The most prominent of these was the growth of the discount houses which rivaled the retailer in many fields. Avalon Hill had distributed exclusively through the retailer system and relied on the salesman to promote their sales. The discount house did not have the salesman-at-the-elbow idea. This meant that a different style of promotion was necessary. Avalon Hill considered making the change as early as 1962 but ultimately decided to wait until 1964. This delay proved to be a mistake.

A second problem arose out of trying to hold the line on prices. The discount house was undercutting the retail prices and games could not be "fair traded" as they once were. This insistence on maintaining the sales price also hurt Avalon Hill.

Thirdly, Avalon Hill had expanded rapidly by drawing heavily on credit. This permitted no serious cash-flow bind. But when, as happened in 1963, dealers representing some 25% of Avalon Hill's sales in 1962 went bankrupt, the cut in the cash flow, plus promotional problems, plus the cut in sales due to holding the prices all combined to cripple Avalon Hill. It could no longer meet its debts.

Roberts had two choices. He could proceed under the Bankruptcy Act, either by way of formal bankruptcy or reorganization, or he could turn the company over to its creditors. A bankruptcy proceeding might salvage the company but would certainly hurt the creditors, who would be forced to accept pennies on the dollar for Avalon Hill's debts. Roberts decided that it would be more honorable to give the creditors a chance to salvage something by keeping the company running. He therefore turned the company over to Monarch Services, one of the two largest creditors. He had paid off the majority of the small creditors and took this course in hopes of giving the remaining creditors a chance to recoup their losses. Roberts then had the option of staying on under Monarch or leaving. He chose to leave, as did Schramm. Shaw chose to stay, and he has remained with Avalon Hill since.

In retrospect, Roberts attributes Avalon Hill's problems to two errors he could have prevented. He could have shifted the promotion system from the "shotgun" approach, selling the whole line, to the "rifle" approach of selling one product and letting it carry the line. This "rifle" technique is the more successful when dealing in the discount system. The failure to make this change when he first saw the need coming was one mistake. The second error was in failing to diversify. Roberts feels that if Avalon Hill had had a second field of operations, such as printing, it would have acted as an "anchor to windward" and actually carried the company over the difficult period in the games.

At this point, Avalon Hill was dead. This is not to say that its products were not being sold. Rather, games were no longer being designed, letters were going unanswered, orders remained unfulfilled. There were still some projects under way which were to have been completed in 1964. There were some games virtually completed and the idea of an Avalon Hill newsletter had been considered in 1963, but it was postponed to 1964 due to growing financial problems. Therefore, while Avalon Hill seemed to be still going quietly along to outsiders, insiders knew that it was dead.

When Monarch took over Avalon Hill, A. Eric Dott (president of Monarch) inspected the assets to see what could be done with it. Dott and Shaw proceeded to go through what was left and Dott concluded that Avalon Hill could be salvaged. He put Shaw in charge, since Shaw was the senior man left over from the Roberts era. Shaw had never done anything outside of the design field under Roberts and had no knowledge of marketing and promotion. As a consequence, Shaw was forced to undergo the same trial and error system that Roberts had encountered six years earlier. His first problem was getting a game ready for the spring toy shows. *Afrika Korps* had been the game Roberts was to produce in the spring, but it was incomplete due to a play problem: the British were able to stop the German advance by putting sacrificial units in the German path. Having no choice but to go with *Afrika Korps,* Shaw was forced to solve the problem. He did so by coming up with the over-run rule and in the space of a few weeks had put the game together. The over-run rule may be Shaw's greatest contribution to wargaming. The failure to solve that problem had held up production of the game for over a year. Moreover, Shaw had to piece together the game from notes left by Schutz when he left. From these notes and with the over-run rule, Shaw started in late December, 1963, when the decision was made to reactivate Avalon Hill, and had the game ready for the toy show in February, 1964. As Shaw points out, those first few months were so hectic that inventory was not taken until July of that year.

The second major move made by Shaw in 1964 was to publish *The General.* Again, Shaw delivered under pressure. The first issue came out in the spring of 1964, in the midst of all of the other tasks necessary to start Avalon Hill running again. *The General* was, and still is, written mainly by hard-core wargamers. In reality, its major purpose was as a house organ to promoted Avalon Hill games, a purpose which has never really changed. When it first came out, however, there was no similar magazine in the field, nor any rival wargame publisher. Therefore, *The General* covered all the major wargames then available. Editorial work was done by Shaw. He continued to perform this function until fairly recently when he turned it over to Randy Reed and Interest Group Baltimore. The latest change is to bring the editing of *The General* back "in house" with Donald Greenwood (founder of *Panzerfaust)* as editor.

As far as Avalon Hill was concerned, *The General* was a stroke of genius. It gave Avalon Hill a ready audience for advance promotion of games. In effect, subscribers paid for the privilege of seeing Avalon Hill advertising. Its second, and more important aspect, was the Opponents Wanted column. For the first time there was a nationally circulated publication that gave wargamers a chance to find out who else was out there.

The dividing line between the time when something is an item of interest to many separate individuals and when it becomes mutually enjoyed as a hobby is usually crossed without anyone knowing it at the time. Roberts has noted that while he then recognized that both a repeat sales market and a vociferous audience existed for wargames, he never considered wargaming to exist as a hobby. As far as he was concerned, at that time, Avalon Hill was an adult games publishing company. Publishing wargames was only a facet of the whole line of games. This is actually a realistic appraisal if viewed from the perspective of 1963. A year later, *The General* had arrived, and that could no longer be said. A hobby was clearly being formed.

Despite the clear evidence presented by *The General's* avid readership, Avalon Hill continued to consider itself an adult game publisher. More importantly, it refused to shift emphasis to wargames in order to exploit the hobby. Avalon Hill *still* remains an adult game publisher.

Roberts never intended to create a hobby and, prior to his leaving Avalon Hill, had no idea he had done so. Second only to the creation of Avalon Hill and the publication of their wargames, *The General* was the most important factor in forming the hobby as it stands today. Despite the venturesome quality of Avalon Hill's actions in 1964, in terms of publishing *The General* and reviving Avalon Hill as a company, the attitude of Avalon Hill became more conservative. The consequences of this were not immediately apparent and are still the subject of debate.

Avalon Hill abandoned the full time design staff. Shaw filled both that job and a number of others by himself. Thereafter, Avalon Hill turned to part-time designers. During the next two years (1964–65), the design work was done during the summer vacation, largely by Lindsley Schutz and Lawrence Pinsky. Between them they designed *Midway* for the fall of 1964, *Battle of the Bulge* for the spring of 1965 (done almost entirely by Pinsky, at home), *Blitzkrieg* for that fall, and *Guadalcanal* for the following spring.

Second, they limited the number of battle titles to two a year: one in the spring and one in the fall. Actually, this works hand in glove with elimination of the design staff. A larger load would be impossible to carry with a limited staff and a large staff would not justify limiting production to two games. The immediate effect was a cost saving of approximately $25,000 for each game not designed. That was the amount tied up in the production of each game. However, aside from the monetary considerations, Avalon Hill believed that the market would not tolerate more than two games a year. They felt that the hobby would become saturated or unable to afford all the games coming out.

This, then, was the basis for Avalon Hill's approach to the following years. The consequences of this policy will be discussed further below.

THE HOBBY IS BORN 1965–1969

Before the creation of *The General,* there was no wargaming hobby. Until then, the history of boardgaming could be neatly compartmentalized and discussed. After *The General,* the hobby crystalized with a speed that few would have predicted. It then grew and diversified at a rapid pace. Avalon Hill never chose to maintain a dominant position in the hobby, as such. The obvious reason for this is that Avalon Hill is in the business of selling games, not creating hobbies. Further, Avalon Hill had no pressure to take the lead. Nobody else put out high quality wargames. If any innovations were to be found, they came from Avalon Hill. There was no incentive for Avalon Hill to be an active leader. Yet, in retrospect, it did not require a great amount of foresight to realize where the hobby would go and what would happen to the hobby if it lacked an active leader. Avalon Hill's benign neglect has been as important as if it had been an active leader.

At Avalon Hill, the conservative program previously mentioned resulted in their turning to outsiders for their games. By 1964, Schutz was an outsider in the sense that he was no longer working full-time for Avalon Hill. Pinsky had never been a full-time employee. These two were the principal designers during the period 1964–65. Shaw, moreover, had never taken the lead in designing wargames. He had been involved in the design of *Blitzkrieg, Afrika Korps, Stalingrad,* and *Midway* more in a supervisory capacity than true design (although his work on *Afrika Korps* was more than merely in passing). It was the wargame *Bismarck* on which Shaw did the most work. Ground combat was not his field, and so others did the design, under his supervision. Playtesting was done during the summer, but there was even less control of the quality of the playtesters than under Roberts. As a result, it is hard to say how much time was wasted by "testers" who were really only around to play the games for fun.

After Schutz and Pinsky, Avalon Hill turned to James F. Dunnigan. Dunnigan had come to Avalon Hill's attention by writing a detailed critique of *Battle of the Bulge,* as well as some detailed monographs on various phases of World War II. He had never designed a game. Why Avalon Hill assumed that he *could* design one, based on his writings alone, is unclear. It probably falls into the same category as the decision to publish *The General*—intelligent guesswork.

Prior to the contracting of Dunnigan, all of the Avalon Hill ground combat games had been a variation of the 1953 *Tactics,* and naval games showed little difference among themselves. Dunnigan believed that different situations required different systems to depict them in a game. Thus, his first two efforts, *Jutland* and *1914,* were unlike anything Avalon Hill had put out. They were both innovative and proved again that there is more than one way to approach a problem. More important to Avalon Hill was the fact that Dunnigan did his work in New York, not Baltimore. The game was designed, in effect, beyond the direct control of Avalon Hill.

Being dissatisfied with Dunnigan's first two games, Avalon Hill turned to Dave Williams, who designed *Anzio.* Williams was as unafraid as Dunnigan in breaking with Avalon Hill tradition, and he too did the design at home rather than in Baltimore. The saga of *Anzio* would fill a book by itself. The "beyond direct control" system encountered real difficulties and Avalon

Hill found it had a tiger by the tail: the game design was not going the way it wanted it to go and Avalon Hill could do nothing about it short of scrapping the game.

The decision to eliminate a permanent design staff and to keep the game production to two a year forced Avalon Hill to seek outside designers. However, in order to avoid chaos in its game design, Avalon Hill then became controlled by its designers since it had to take them where it found them or forego their work. If Dunnigan did not want to move to Baltimore, they either put up with him working in New York or lost his services. Another example of decentralization was the answering of letters. Initially, Roberts took pride in a formal written response to each letter. When outside designers became the practice, they were obligated to answer all inquiries on game play for a certain period. Someone else was given the job of responding to letters on the older games. This person might be as far away from Baltimore as Albany, New York. At that distance, Avalon Hill had no real control over what was said to the writers. Interest Group Baltimore acquired the job in recent years, bringing it back "home." During this time, however, it was quite beyond Avalon Hill's practical control.

As mentioned above, the growth of the hobby became less and less orderly as time went on. Several changes happened at once. While Avalon Hill was decentralizing, another facet of the hobby was forming: the clubs. The majority of the hobbyists are not club members, but the clubs—being the most organized—made the most noise and had an influence far beyond their size. Oddly, the low profile Avalon Hill took prevented the clubs from having the impact that they might have had. Avalon Hill simply could not react to the clubs with any real flexibility.

A third element, growing at the same time everything else was going on, was the hobby-produced magazine. Few were started with any intent to rival *The General*. Most were spiritmaster or mimeographed. The term "zine" has often been applied to them. That term, in turn, is taken from the world of science-fiction fandom, and anyone who is familiar with a "fanzine" knows why the term applies. There were few pretensions to quality at first. Often as not, they were club organs or publications with very small circulations (a few hundred subscribers at most). They usually did nothing more then give the editor/publisher/sole author a chance to voice his opinions. Further, they came and went with frightening regularity, usually taking with them any subscription payments that had not already been spent. 'Zines, like the clubs, were the creation of *The General*. This is not in the sense of mimicry, but because *The General* gave free advertising and gave the 'zines a purpose—there is no need for a newsletter if everyone lives in the same town, but statewide or national circulation is another matter.

At this point, it is fitting to note the arrival of *Strategy & Tactics*. Christopher R. Wagner, then a Staff Sergeant stationed with the U.S. Air Force in Japan (and environs) had decided that something more could be done with a wargame journal than *The General* was doing. First, he saw no real signs of "edited" articles in *The General:* as long as articles were not libelous, they were published. Second, he saw no cohesive editorial concept holding the magazine on one course or another. Third, the quality of the articles was very uneven. The few good writers were counterbalanced by the mediocrities. At the same time, he did not want to put out another 'zine, with poor reproduction and the reliability of mist.

As a consequence, in 1966 he and Lyle E. Smethers, another fellow stationed with him in Japan, decided to put out their own effort. He wanted an alternative to *The General*. Wagner decided to add areas excluded from *The General,* such as miniatures and, as they appeared, non-Avalon Hill wargames. He contacted Shaw who was helpful but mildly discouraging. In this instance, Wagner agrees that the discouragement was not that given to a potential rival but sound advice. Wagner calculated that he could achieve fifty percent of *The General's* circulation. Shaw had indicated that *The General* had a circulation of 5,000. If he had a circulation of 2,500, *S&T* would be successful. Thus, he started off with a printed magazine on glossy paper. His fifty percent estimate proved correct. *S&T* did achieve fifty percent of *The General's* circulation, but the *paid* circulation of *The General* was only around 2,500. With a circulation of 1,200, *S&T* was doomed to be economically unsuccessful.

Wagner got the magazine off the ground by soliciting articles through the mail. Since he was in Japan, he got Henry Bodenstedt, a toy shop owner in Adelphia, New Jersey, who was interested in miniatures, to write articles about miniatures and distribute the magazine. In exchange for this, Bodenstedt got free advertising and thus advertised items which were, curiously enough, ideally suited for use in the games he was writing about in *S&T*.

Wagner ran the magazine from Japan, getting articles from the United States, negotiating with the Japanese printers, sending the finished copies to Bodenstedt and starting the cycle again. His initial band of contributors came from three areas. The first group were his friends in the Air Force such as Smethers and Scott Berschig. Second, he used some of the people who had written for *The General,* such as Donald Greenwood, Louis Zocchi, and John Dotson. Third, there were those such as Dunnigan who came to Wagner's attention through published monographs.

The parting of the ways between Bodenstedt and Wagner is more fully detailed in *S&T*'s reprint of issues 1 through 6, now available as *Book I.* Suffice it to say that Wagner subsequently used his mother to distribute the magazine. At the same time, he decided to leave the Air Force and return to the United States. Issues 1 through 10, constituting the first year's publications, were all printed in Japan. Thereafter, they were printed in the United States.

During the first year, *S&T* continued to grow in quality and quantity. Illustrations became better and a glossy cover was first used for issue 7. In addition, paid advertising appeared in the magazine. The increased quality was not coupled with a serious effort to promote new readers, however, and the overhead began to grow. Returning to the United States, Wagner found the cost of printing the magazine to be much greater than it had been in Japan.

During with period, the editorial staff underwent some upheaval. Many of the original writers lost interest. However, some regulars, such as Rodney Walker, began to appear. In the middle of the second year, Wagner secured the services of Redmond A. Simonsen, under whose artistic hand the over-all appearance of the magazine took a major jump.

By the second year, the story of *S&T,* from the inside, was a quest for money. Wagner had been trying to make a living from *S&T* and had even paid some money to contributors. The end of the first year, however, saw a renewal rate far below the 100% he had expected—closer to 70%, in fact. Without a comparable lot of new subscribers, the circulation began to decline. Wagner had exhausted the sources of subscribers available to him which meant that *S&T*'s circulation slowly spiraled down. He tried to sell games through *S&T* in hopes of helping the magazine with the profits from the sales. Notable among the lines carried was Gamescience from Philip Orbanes. But this leveled off so that by the time issue 12 came out, Wagner took a part-time job and began to borrow money to keep the magazine afloat. His problems were further compounded by printer delays. On one occasion, he personally collated the magazine late at night in the printer's shop. He changed printers twice in the last two issues under his administration, but to no avail.

Despite financial problems, he did not let the quality slip. In fact, the first two-color issue was the last under Wagner—Volume 3, Number 1 (issue 17). But, as Wagner acknowledges, reality has a way of catching up. By the time issue 16 was on the presses, he knew he could either print number 17 and leave the printers with uncollectable debts or stop at 16 and leave the subscribers hanging. He printed number 17.

At the same time, he put out feelers to see if anyone would take over the magazine. The two major wargaming clubs, Spartan International and International Federation of Wargamers, were both approached. Things reached the stage with IFW that some materials were actually shipped before the deal fell through. *S&T* had reached the end of its tether. No one wanted it. Wagner could not afford it and Simonsen, *S&T*'s premier artist, was disillusioned to the point of having no further interest in it.

Still another facet of the hobby growing during this period was the non-Avalon Hill, non-major company game. The first efforts were poor, to say the least. They were often spirit mastered with sheets that never matched up and counters that disintegrated on touch. It was apparent, before long, that this type of game would have only marginal success. The question was whether a well done game could have any success outside of Avalon Hill.

At this point, Philip Orbanes becomes the focus of attention. Orbanes had ventured into the game design field with a spirit-duplicated game called *Operation Gigantus* in 1964, while still a high school student. The game was a cross between *Diplomacy* and an Avalon Hill game. Although this met with only moderate success, Orbanes decided to go further. He formed Gamescience, a corporation, in 1965 and published a game called *Viet Nam*. This was a high quality game with a regular mounted game board, plastic pieces, and a white box with a label. This was sold by arrangement with Avalon Hill in that Avalon Hill operated as a funnel for sales by accepting orders which they would then forward to Orbanes to be filled.

In 1967 Orbanes produced *Confrontation*, a game printed on heavy cardboard in three sections and mailed in a large tube. The map was an overprinted Rand-McNally world map which formed a gameboard sixty by thirty inches. This system of packaging was so successful that Orbanes re-worked *Viet Nam* along the same lines. This game was distributed in same manner as *Viet Nam*.

Orbanes' real breakthrough came as the result of Avalon Hill rejecting Louis Zocchi's game, *Battle of Britain*. Wagner was aware of Zocchi's game and believed that the title was a good one. Wagner suggested that Zocchi and Orbanes get together with the result that *Battle of Britain* was a major success.

It proved so successful that Avalon Hill stopped helping Orbanes' sales and treated Gamescience as a rival.

It was not a rival at that point. Whether it might have been is a moot question. Orbanes decided to invest a major amount of money in the game and sell it at the February toy show in Chicago. In the interim, he upgraded *Viet Nam* and *Confrontation* by printing the maps on Texoprint (something like oilcloth, only better), with die cut counters and printed rules. *Battle of Britain*, on the other hand, was in the then standard Avalon Hill format of a long, flat box. At the toy show, Orbanes was approached by a representative of Renwal, the model manufacturer. They wanted to expand their line into other toy fields and felt *Battle of Britain* to be the game. As a result, they persuaded Orbanes to sell all his assets to them in exchange for royalties. He kept the company's liabilities (which were rather large by then, due to promotion expenses) but expected to pay them off soon from royalties. In fact, royalties never came in any large amount. Renwal sold only a fraction of the amount Orbanes was selling earlier, and thus ended the first rival to Avalon Hill.

What Gamescience brought to the field was the concept of different systems for different situations. This was the same idea Dunnigan had explored in *1914* and *Jutland*. Moreover, the success of *Battle of Britain*, after it had been rejected by Avalon Hill, gave dramatic proof that Avalon Hill could be wrong.

Actually, Gamescience did not disappear. The name was taken over by Renwal; the corporate shell (and debts) remained. Orbanes and

Dunnigan decided to use the corporate shell and form Operation Design Corporation which, in turn, became a subsidiary of Infinity Quest. The function of ODC was to provide a corporate shelter for Dunnigan's winsome creation, Poultron Press which, in turn, was the new publisher of *Strategy & Tactics*.

The story of the salvaging of *S&T* forms a suitable end to this period, as well as marking the start of the current phase of development.

As mentioned, Wagner had been casting about for someone who would take over *S&T*. Dunnigan had been approached but saw it as too big a job. However, Dunnigan had been working with the idea of producing a large number of games. Obviously he could not sell them through *The General* since Avalon Hill was not likely to cut its own throat. *S&T* was the only other magazine with any circulation worth talking about. The net result was that Dunnigan decided to take over *S&T*. He had a group of friends in the New York area interested in the games and he grabbed the most visible of them and told them that they were going into the publishing business. He also contacted Redmond Simonsen to do the art work. Simonsen, in essence, told Dunnigan that he had had enough of *S&T* after the work done on issue 17. As a result, Dunnigan set about creating issue 18 out of whole cloth. A few articles from the regular writers were available, but the great body of issue 18 was written and pasted-up by Dunnigan. When Simonsen saw that issue 18 was actually going to come out, he agreed to do a quick cover and some incidental graphics and thereby began to re-involve himself with *S&T*. With that melodramatic turn of events, issue 18 was printed; and thus *S&T* was salvaged. More importantly, Dunnigan introduced several games in issue 18 which were to be the core of the Test Series Games, including *Tactical Game #3,* the model for *PanzerBlitz*.

THE SECOND GENERATION 1969–1972

The period after 1969 saw the birth of the second generation of virtually all aspects of the hobby. The warring clubs of the previous period were replaced by the more mature clubs such as Spartan International, AHIKS, and IFW. Even the 'zines took on a better quality. Club newsletters and the like lost most of their parochial tendencies and turned toward general interest. At the same time, spirit master and mimeograph were replaced by offset for the 'zines in more and more instances. Those ' zines also produced a group of writers whose works began to appear in the better quality 'zines with increasing regularity. Some of the old 'zines from the first generation still lingered. They tended to suffer from wordiness, at best, and a lack of critical insight, being nothing more than thinly disguised vehicles for the views of their editors. However, considering that the 'zines were produced by people working on a part-time basis with limited budgets, most of these publications probably reached their maximum quality potential.

During this period, *S&T* underwent a major change in direction, tending steadily away from a pot-pourri of articles into a format emphasizing history. The magazine became, first and foremost, a history magazine. Gone were the variants, and hot on their departing heels were the articles on miniatures. Dunnigan introduced the feedback to monitor reader response, and it became a self-sustaining process: the change in emphasis attracted people who liked the new format and who, in turn, responded with low ratings to non-historical and nonboard game articles. *Diplomacy* was the most tenacious of the pre-Dunnigan era features, retaining a small but loyal constituency. But that, too, was at length eliminated, and *S&T* moved into its present mold. During this time, *S&T* exercised an increasing influence on the hobby. Whereas prior to 1969, 'zines emphasized "perfect" plans and variants, they now invested more space in military history and the like. Articles became compatible with, rather than subordinate to, the game. Seeing the success of the magazine games that *S&T* began to publish, other companies published games. Initially, the shift in orientation may have been premature in terms of retaining the old subscribers to *S&T*. Many readers simply did not like the new format and failed to re-subscribe. There was a period when *S&T*, like Carroll's Red Queen, was running as fast as it could simply to stay in the same place in its circulation.

The second major impact that *S&T* and Poultron Press/SPI came to have on the hobby was its introduction of a flood of games. Avalon Hill had long since espoused the theory that the wargame market had a low saturation point and that more than a few games per year would glut the market. They also felt that the number of suitable subjects was limited. Dunnigan disputed this theory and set out to illustrate its invalidity. First he introduced a great number of games through *S&T* and the Test Series Games. Almost at once, the number of games on the market (i.e., those released by Avalon Hill) was doubled. Second, the pat formula of game design that had characterized the majority of the Avalon Hill games was abandoned. Whereas most of the Avalon Hill games were such that the rules for one could be applied to most others, Dunnigan developed different game systems for various situations. While some concepts resulted in a series of games—games based on the same system and unit level such as *Kursk, Turning Point, France '40,* and *Moscow Campaign*—often as not a given game system would be no more than a distant cousin of a predecessor.

This flood of games had a major impact on the hobby. If only in response to the novelty, gamers began to look forward to the *S&T*/SPI game to see where the company would go next. The attempt to achieve some precision in rules writing led to deliberately-coined terms. Because there was no comparable jargon elsewhere, people began to adopt the new terms into general usage (with a few exceptions, as in the persistence of "square" to describe hexagon). In part, this terminology gained "official" approval when Avalon Hill, as part of its outside design policy, had SPI design *PanzerBlitz* and *France '40* for them, thus permitting the interjection of SPI terms into Avalon Hill games.

At this time Avalon Hill seemed caught in a period of doldrums. It chose not to have in-house designers, preferring to go to outside sources for its games. The sole major exception to this tradition was *Kriegspiel*. Though apparently successful economically, *Kriegspiel* was a critical failure among wargamers. AH's major reason in publishing the game was to venture into the bookcase-type of game in order to determine wargames' suitability for that format. It certainly proved their point, even if the game was a critical failure, and Avalon Hill has stayed with the bookcase format for all of their new games, with few exceptions, since that time. *Kriegspiel* was the only wargame designed by Tom Shaw, which is ironic since he did have a hand in some of the greater Avalon Hill successes—notably *Afrika Korps*—and he deserves more credit from wargamers than that brought to him by *Kriegspiel*. A major innovation in *Kriegspiel* was the adaptation of the matrix combat results table from the AH series of sports games. This was the first wargame not to require dice to decide combat.

When *Kriegspiel* failed among hobbyists, Avalon Hill found itself in a bind of its own creation. AH did not have the in-house manpower to design the more complex games that had characterized the company's earlier production—games of the type that SPI was producing and the public was demanding.

Because they lacked the staff, they no longer had the option of doing games as they wanted. They were forced to resort more and more to outside designers. For one man to do a good game takes a considerable amount of time, whereas the teams that SPI was establishing could do the job in a much shorter time. *PanzerBlitz, Luftwaffe, Origins of World War II,* and *Outdoor Survival* not only came from an outside designer, but they all came specifically from the *S&T*/TSG/SPI complex. *Luftwaffe* had been designed by Louis Zocchi, but it was already published in an earlier version as *12*

O'Clock High by TSG. *PanzerBlitz,* under the name of Tactical Game 3, was also a TSG creation and *France '40* was published in *S&T*. Only *Origins of World War II* and *Outdoor Survival* were totally new.

THE RIVALS 1972–1976

In 1972, and 1973, SPI caught fire and began to grow by leaps and bounds. The circulation of *S&T* surpassed that of *The General* and became the most influential magazine in the hobby. More importantly, the quality of the games produced by SPI improved both physically and technically. This upsurge in popularity had its drawbacks, however. There were those who felt that SPI had outgrown its role as the industry's upstart and had become part of—or perhaps all of—the establishment. Those who were disenchanted looked elsewhere, and into this breech stepped a number of new companies.

The greatest phenomena of this three year period is the number of serious efforts that were made to establish rivals to SPI and Avalon Hill. Perhaps the most substantial effort was that of Simulations Design Corporation, a company that produced both separate games and a magazine entitled *Conflict*. *Conflict* was a *tour de force* that featured a glossy format and four color illustrations right from its inception. The magazine rapidly attracted a small group of fans who felt that SPI had grown too large and that AH had also let them down. SDC, as a small company, was faced with the age-old problem of an appetite bigger than its wallet. It simply could not sustain such an effort on such a small circulation. It had eventually to grow rapidly, constrict its operations, or go under. *Conflict* chose to forego the first two options, even though it covered essentially the same ground as —and therefore competed more or less directly with—*S&T,* with articles on history, game reviews, and game publication. Finally, financial reality caught up. The lag between issues got longer and at length, with issue 8, their printer demanded cash on the barrelhead, and SDC had to suspend publication. This is not to say the *Conflict* is gone forever. SDC still exists as a company and still sells games on a limited basis. Should the company ever inherit a large sum of money, *Conflict* could well reappear. But for the time being, the best imitation of *S&T* is gone.

Panzerfaust posed a more interesting problem. When Don Greenwood, the original publisher, turned the magazine over to Don Lowry, *Panzerfaust* was the fourth largest 'zine in the hobby behind *Moves, The General,* and *S&T* and well ahead of the club 'zines of IFW and Sparta. Lowry also had his own line of games he was interested in publishing, Guidon Games. Mounted on boards with full color maps, these games were physically superior to those put out by SPI, although they suffered from obscure rules and mediocre graphic skill. They were never any great success. What happened to *Panzerfaust* is, as of this writing, one of life's little secrets. For a brief time, nothing was sent to the loyal readership. The magazine simply stopped publishing and nothing was heard of it, or the games, until the latter part of 1975 when Don Lowry, reappeared in California with a magazine he called *Campaign,* subtitled *Panzerfaust,* with an obviously smaller circulation than that of the original *Panzerfaust*. *Panzerfaust/Campaign* was a different piece of business from *S&T, Conflict,* or even *The General*. There were historical articles on the *S&T* model, but they lacked the real depth and capacity to supplement the games that SPI or SDC put into their magazines. *Panzerfaust/Campaign* did not opt for a game in each issue but chose to serve more as a vehicle for criticism of games from various companies. Its special *bête noir* has been SPI, and letters to the editor columns are filled with pro and con debates as to whether Lowry's particular bent is valid or not. Parenthetically, since the period of low profile, nothing has been heard of Guidon Games.

With SPI and Avalon Hill aiming their product at a wide market, there was room for a company which specialized in games for the buff who was willing to step into highly complex weekend-long games. Game Designer's Workshop filled that gap. Working on the premise that the games in which they were interested would probably interest other people too, GDW has produced a series of simulations on rather unusual and obscure subjects, such as Drusus' campaign to recover the legionary eagles lost when Varus was defeated in 9 AD in the Teutoberger Forest (*Eagles!*) and the Bolivian-Paraguayan war in the Gran Chacco desert in the 1930's (*Chaco*). However, their real claim to fame has been their critically acclaimed Europa Series—a series of games that, when linked together, would form all of European World War II. The origins of this series can probably be traced to a time well before GDW ever thought of forming a company. There had been several gigantic maps of Europe on which people sought to recreate the whole of WWII in Europe, using components of old Avalon Hill games. SPI had been threatening to produce a "Stalingrad II' game that would be a giant Eastern Front game. GDW then conceived a same-scale series of games on all of the aspects of the war in Europe. To date, they have published two games on the Eastern Front, a game on the invasion of Norway and Denmark, and an invasion of Britain game, with a 1941 Balkan campaign game in the offing. Critics have pointed out that the rules are not as clear as might be desired, but the myriad GDW fans seem untroubled by this situation. Because of the complexity of all the Europa games, GDW has garnered a distinct following from among the hard information buffs, and from these gamers GDW receives exceptionally high marks in game-rating surveys. On the other hand, GDW games are not for everyone because of that same complexity. Still, their following is very loyal and, with the acquisition of John Hill's Conflict Game Company, GDW has moved into the third position in game publishing—albeit a distant third—behind SPI and Avalon Hill.

John Hill is an interesting phenomenon in the hobby. Running a successful hobby shop, he was not forced to look to games as a source of income as he created them. Rather, he could design them as he saw fit to satisfy his own interest. Using an almost conversational style of rules writing, Hill turned out a series of games for his own Conflict Game Company and for SDC that included *Hue, Verdun,* and *Kasserine Pass*. He acquired a loyal following among those in the hobby who could cope with his open-ended rules, and he soon found that he had created a Frankenstein's monster—a company that made increasing demands on his time while providing only minimal income. In 1975, he sold his company to GDW, although he still remains active in game design. As a one-man game-design company he is, to this point, unique in the hobby by virtue of the volume of his output and the originality of his ideas. As such, he is one of the few individual designers recognized by name in the hobby.

The end of this period saw a proliferation of game designers (Battleline, Jagdpanther, TSR, etc.) and games. One major addition was the science fiction game. Originally published by what could be characterized only as "fringe" and "one-shot" companies, these games appealed directly to the most common connection among wargamers: more wargamers have an interest in science fiction as a second hobby than anything else. When the science fiction games first appeared, they were unusual in that none of the conventional wargames had really dealt with warfare of even the near future, much less the distant future. Only in a very few games was it possible to play a futuristic game and that, with one or two exceptions, was usually accomplished by means of a variant of a basic scenario of another game. The science fiction games, with wars in this solar system and in other galaxies, opened up a new field and showed that the hobby would accept ahistorical games. In fact, SPI feedback shows that interest has shifted from World War II to the present and the future eras, a fact that is reflected in the growing number of SPI games devoted to events which have not happened and may never happen.

By now, Avalon Hill was faced with the unpleasant fact that their position as sole proprietor—or even leader—of the hobby was being challenged by SPI. Consequently, they began to hire their own full-time design staff for the first time since the end of the Roberts era. First, they got Don Greenwood who had made *Panzerfaust* into a credible effort. His impact was first felt in *The General*. That venerable journal began to publish articles other than the umpteenth variant on *Stalingrad* and the quality of the entire magazine improved considerably. The second step AH took was to hire Randy Reed, one of the leading figures in Interest Group Baltimore, as a full-time member of the R&D staff. Reed's first effort as a game designer was *Richthofen's War*, and the game's success, compounded by the fact that it was produced by Avalon Hill, made Reed an instant figure of importance in wargame design. However, Avalon Hill still chose not to establish a fully in-house design staff. As a result, Don Greenwood had to spend many hours reworking *Third Reich* into something that met Avalon Hill's standards after that game had been designed by an outsider.

This dichotomy between internal and external design has continued at Avalon Hill. Some games have been done internally, such as *1776*. Other games are reworked versions of older titles, such as the revised *Anzio*. Still other games are totally outside efforts, as in the case of *Tobruk*, or redone games of other publishers, such as *Wooden Ships and Iron Men*. The majority of the games produced by AH in recent years continues to be based on outside designs. Avalon Hill uses its own forces more in the development stage. Avalon Hill has undergone a major change in direction and in degree of innovation with the result that the company may well have preserved its position at the last minute. Prior to these changes, protestations to the contrary notwithstanding, gamers were looking to Avalon Hill more out of habit than from the belief that something new or better would be forthcoming. To a large extent, this situation has changed, and the regeneration of Avalon Hill as a major publisher of wargames is an occurrence of much promise to the industry.

The period since 1972 has been, if anything, the era of the hydra. Like that mythical beast, when one company falls by the wayside, several more appear in its place. One of the more interesting of these companies is Rand Game Associates. RGA resembles, in its staff, an SPI alumni association. Many of its staffers have worked for SPI in one capacity or another at some time. Perhaps because of this proximity, RGA has tried deliberately—and sometimes comically—to avoid similarities between their games and those of SPI or Avalon Hill. One of the more interesting innovations of RGA was their command series games. The theory was that a person would, for a flat fee, subscribe to a series of games without any magazine or back-up material. Subscribers received a box to hold the games and some standard components. The rest would be received upon publication. The first year's subscription went well enough, and some of the games—like *Cambrai*—were well received. The time schedule was more or less adhered to, and if there was any dispute as to the quality of the games, people assumed that the second year, building on the experience of the first, would be better. Instead, the second year was a disaster. Production fell rapidly behind schedule, and fewer than half of the proposed games were published. Moreover, RGA shifted to producing larger scale games in addition to the command series. While these achieved a greater degree of critical success, they were also produced by designers outside the RGA formal staff. In mid-1976, ownership of Rand Games changed hands, leaving the future of RGA in question.

One of the more interesting developments of this period were the non-board wargames. These games required an umpire or moderator who, in some games could play an active role, while in others he was simply the person who held things together. Notable in this field was Gary Gygax's *Dungeons and Dragons* (TSR); Flying Buffalo's *Nuclear Destruction* and *White Bear, Red Moon* are also quite popular. All have the dual advantage of permitting many players to play, something usually not possible with conventional wargames, and permitting each session to be radically different from the one which preceeded it. Conventional wargames, based on a set history or premise, restrict the number of possible variations. Thus, these games, and the types of games which have spun off from them, have attracted a certain following that rises almost to the level of acult.

Another cult game emerged in mid-1975, and this proved to be something of a surprise. In the first place, the subject was quite obscure (the Wars of the Roses) and, additionally, the publisher was English. Few, if any, games published outside the U.S. had ever had any impact on the wargame market. But *Kingmaker*, a game that loosely followed the style of the venerable *Diplomacy*, proved to be a smashing success. Initially imported by SPI, the game was bought *in toto* by Avalon Hill in 1976, and its popularity has risen sharply to the point where it is now one of the highest-rated games in the field.

Cult games are unusual (only *Diplomacy* has withstood the test of time as far as this is concerned) and no one game has ever assumed the importance and popularity of a *Monopoly* or a *Scrabble*. The problem seems to be more of consumer unawareness and even resistance, for while the scope and reach of wargames has, in the past four years or so, grown in incredible leaps, it still remains a pastime unknown to the majority of the populace.

THE MEN BEHIND THE GAMES

Charles S. Roberts The Founding Father

It is fitting that a hobby appealing to people generally in their twenties should be started by a man of the same age. If anyone can be called the father of the hobby, it would be Roberts who, while in his early twenties, conceived *Tactics* and before he reached thirty had founded Avalon Hill. As a consequence, now in his early forties he is scarcely in his dotage and might smile to be considered the father of anything except children.

The early growth of Avalon Hill reflects Roberts' personality—outgoing and assertive. So, too, its failures reflect his faults. He is frank to admit that procrastination in shifting to single product marketing was a serious mistake. Activity was coupled with periods of inertia. However, the one comment constantly made about his administration of Avalon Hill is "integrity." Nowhere is that more evident than in his decision to sell out to his creditors rather than leave them in the lurch via bankruptcy.

The emphasis of the company was also reflective of Roberts. He had been interested in World War II, notably the Russian front (and in fact has a desire to write a purely military history of that part of the war), and this interest is reflected in the line of wargames, the single largest part of the Avalon Hill line while he was there. His insistence on a broad spectrum company is also obvious and, in fact, his own personal favorite among all the games he designed is *Management*.

Perhaps the greatest irony, however, is best expressed in Roberts' own words: " . . . what exists now, which we created at Avalon Hill in those early years, not so much by design but by default, it's the hard core of a new hobby. . . . That was not our intention, though . . . certainly wasn't mine."

Nevertheless, Roberts is pleased with the growth of his creation. Even before he left he recognized that there was something there, as evidenced by tentative plans for a newsletter, later to appear as *The General*. The only development which he found displeasing was the emphasis given in certain quarters to the Nazi-/German aspect of the hobby. He recognizes that World War II is the most dramatic and fertile field for wargames and that the Germans do present both a challenge and a "colorful" side to consider. Nontheless, he rejects the idea that this should lead to the almost neo-Nazi activities, found in some clubs' names and tactics, as they existed in the not-too-distant past.

Roberts still lives in the Baltimore area, now working as an executive in a printing firm. Though his time in the wargame field totalled no more than ten years, he began something the end of which is not in sight.

Thomas N. Shaw
The Man In The Middle

Thomas Shaw is a direct contrast to Roberts both physically and temperamentally. Shaw is a rather quiet, methodical person as opposed to Roberts' outgoing, almost mercurial approach. Although Shaw and Roberts had gone to high school together, the sole reason for his joining Avalon Hill was his design skill. In that function he controlled the Avalon Hill Design team but had nothing to do with marketing or management. However, when the Roberts era ended, Shaw was put into all of those jobs by A. Eric Dott. As a result, while Dott has the final decision, Shaw puts it all together and makes recommendations. As a result, he is truly the man in the middle at Avalon Hill. He is the most visible one there—many people are unaware that he is not the head of Avalon Hill. Until recently, he edited *The General* himself and, particularly since Pinsky and Schutz left, he has kept open the lines of communication with the hobby and game designers. Had he not, Avalon Hill may well have left the wargame field.

Although a game designer at heart, Shaw does not consider himself a *wargame* designer and is only involved when he must be. He provided continuity from the Roberts era with the result that there was little visible change in the type of product produced (particularly in the wargame field) until fairly recently. In that vein, he has ensured that Avalon Hill remains active in the field of adult games.

As the man in the middle, it is easy to fix upon him when discussing the basic Avalon Hill philosophy toward wargames as he is their most visible advocate. Shaw has expressed the idea that wargames, while providing a constant market of about 20,000, will never provide the big sales. Avalon Hill's records bear this out, at least as far as their games are concerned. On the other hand, Avalon Hill has never tried to see how hard they can push the market. In all fairness to Shaw, he cannot compel Avalon Hill to expand its sales in wargames by an intensive sales campaign. That would be Dott's province. On the other hand, only Shaw, as the man in the middle, seems capable of putting together the information which might change Dott's mind.

Christopher R. Wagner
The Man and The Hobby

Christopher Wagner is important to the hobby because he founded *S&T*. Wagner had become interested in the hobby in the 1960's. While a member of the Air Force, he was sent to Yale to learn Chinese. Although the press of military activities and the initial absence of any opponents caused him to lose some interest in the hobby during the next few years, after a while more men were assigned to his unit in Japan and he found others with the same interest.

The reason why Wagner created *S&T* is mentioned elsewhere. The mere fact that he did so is of importance. *The General* was then, as now, a house organ of Avalon Hill and primarily concerned with making money by promoting Avalon Hill products. Wagner's determination to put out *S&T* in professionally printed form required a basic determination which no one had had the opportunity to make before: that there was a hobby and that it was big enough to afford a living to someone (or a group of people). That his efforts proved unsuccessful, in that he ran into financial trouble, is immaterial. His error was in assuming that the hobby was large enough in 1966 to sustain a magazine such as *S&T*. In fact, it was a very small hobby. It had grown almost exclusively by word of mouth. Avalon Hill was only interested in promoting its games, not the hobby (which is as it should be for a company in the business of making money). Without any real effort to cause the hobby to grow, progress could only be slow.

S&T was therefore doomed before it started. In point of fact, *S&T* can still not sustain its level of quality on subscriptions and advertisement alone, even though its circulation is twelve times what Wagner had at his best.

Wagner has noted that if he had known what he came to find out the hard way, he never would have started *S&T*. Had he not, someone else would have, because the hobby needed such a magazine. If another magazine had not been started, the club magazines would probably have filled the gap, though perhaps not as rapidly as they have done with *S&T* leading the way. The need was there in 1966, but the hobby was not big enough to sustain *S&T*. The decision to try it despite such problems is Wagner's great contribution to the hobby.

James F. Dunnigan
The Dean of Modern Wargaming

James Dunnigan is an unlikely person to occupy the position of "Dean of Wargame Designers." He is, in fact, a high school dropout, a point he delights in making to those who inquire as to the prerequisites of being a wargame designer. As with Wagner, Dunnigan became interested in the games at the start of his military "career"' (1961-64), lost interest after he went overseas, and then became interested again. Actually, he determined, while in the service, that a college degree would be a good thing to have so he remedied the high school deficiency and got his degree from Columbia. It was while a student at Columbia that he first came to the attention of Avalon Hill. He had written a detailed critique of the *Battle of the Bulge* and from this Avalon Hill reached the non-sequitur conclusion that he could also design games. As he had never done so, this appeared a novel idea and Dunnigan tried it. The result was *Jutland*.

At the same time, he produced a series of pamphlets on various World War II campaigns. These pamphlets brought him to Wagner's attention and Dunnigan was involved with *S&T* on a limited basis from the start.

Dunnigan's contributions to the field stem from the basic idea that nothing should be taken more seriously than it deserves to be. This, coupled with the belief that history is a great deal more than a cold recitation of facts, allows him to make unorthodox appraisals of history. When applying these to wargame design, he developed the premise that different situations require different solutions and can be depicted in a game through a "systems analysis" approach. The result was, for the first time, a reasonably accurate simulation of a military/historical event. It is this same premise that appears again and again in games he designs or which are designed under his supervision.

It is always easier to use a tried and true format and make minor alterations to suit the situation. That was the old Avalon Hill system. The Dunnigan system assumes that nothing that was used last time will necessarily be used this time. If it can be used, fine. If not, don't hesitate to throw it out and find something more suitable.

Out of the same basic philosophy is the flood of games produced by Dunnigan and others under his direction. Whereas Avalon Hill has taken the position that the market could be glutted by too many games, Dunnigan believes that the more games there are, the better the choice. Given the right subject, anyone will buy one more game. Avalon Hill contends that the number of topics for games is limited. Dunnigan disputes this by reeling off a list of potential titles.

The result of the seemingly unerring Dunnigan touch (though he will readily admit to having made mistakes) is that people tend to speak of him as a genius, a notion he rejects as invalid. Actually, in this instance the key to his "genius," if the term is to be used, is readily recognizable. It is his refusal to be bound by a fixed pattern when that pattern does not fit. We are all creatures of habit since it is easier to do the same thing until you achieve perfection than try to do a task a different way each time. Routine is a common practice. To develop a new method of doing something always requires more effort than following a pattern. So, while others are following the pattern set out in the last game, Dunnigan is off developing a new system. *The General* works on the theory that a given article will generate a number of publishable replies which will in turn spawn new replies and give a "safe" source of articles for issues to come. Dunnigan, on the other hand, will not hesitate to try something just because it seems interesting and has not been done before. If it succeeds, all well and good. And if it fails, then at least a lesson has been learned.

If there is any irony in it at all, it is that Dunnigan rarely plays the games he has designed, or any others for that matter. The intellectual satisfaction of designing has replaced that of playing.

Redmond A. Simonsen Graphics and Gamespeak

The most subtle and most pervasive touch in all of Simulations efforts is that of Redmond Simonsen. Subtle because the nature of his work is all but unnoticed until one tries to duplicate it; pervasive because the final form and appearance of *S&T* magazine bears his touch.

Simonsen is a trained artist (BFA, Cooper Union, 1964) who did make a living at the field of graphics prior to working for SPI on a full-time basis. As such, he is a unique find. Moreover, of all the people at the top of the heap in SPI, he probably plays the games the most. The combination of artistic ability and an interest in games is obviously the best thing that happened to *S&T* and wargames in general since *The General*.

In fact, Simonsen sought *S&T* out, rather than the other way around. He subscribed back at the time Wagner was running the magazine, saw that he liked the magazine and determined that the graphics left much to be desired. Almost as a personal challenge, he decided to draw some illustrations for a hypothetical magazine (this without even talking to Wagner on the subject) to include a cover, interior illustrations, and the logo. He submitted the package to Wagner who, recognizing a good thing, grabbed what he got. The now familiar *S&T* logo was first used on issue 13, together with other of the illustrations Simonsen had submitted. The experimental cover he had designed eventually appeared in issue 16.

From that point on, Simonsen became *S&T's* artist. He worked out much of his contributions by mail and infrequent visits between his home in New York City and Wagner's in Albany. But despite the distance, his influence became more apparent. His efforts culminated in the two-color issue, number 17, that ended the Wagner era. As mentioned elsewhere, he became disillusioned and at first rejected Dunnigan's offer to rejoin *S&T*. In fact, little of issues 18 and 19 were illustrated or designed by Simonsen. But thereafter, he took a more active hand and has since carved out a permanent niche.

Simonsen's contributions in the field of graphics are, as mentioned above, difficult to recognize unless pointed out. Then they become readily apparent. The very layout of the magazine is his determination, in terms of illustrations, type styling, and general blocking of articles. More importantly, his skills in graphics are of major value in the games produced by SPI. His most significant effort has been the attempt to minimize the number of pieces of paper one must use to play a game. He has done this by transferring data to the game map and, more importantly, to the counters themselves. The ever-increasing numbers on the counters all reflect the desire to provide the player with the maximum amount of data available at his finger tips, rather than buried deep in the rules. This, in turn, allows more complex rules since the piece contains the information necessary to apply the given rules.

The second area of interest where Simonsen's influence is felt is in the rules themselves. He has always rejected the notion that there need be any confusion in the rules for games, even if they are complex. He has therefore tried to standardize the terms and the formats used in the game rules and make the terms bear some relation to the situation they depict so that they are as self-explanatory as possible. In addition, Simonsen's talents as a game designer, especially of the popular *StarForce* and *Sorcerer* games, can be assessed in the symmetry, balance and novelty inherent in the systems.

Simonsen's role is not a readily recognizable one, yet it is one which gives SPI products a unique flavor as surely as any author's style and has already had a major impact on the hobby.

Randall C. Reed

The nice thing about working for Avalon Hill is that it affords an opportunity for instant visibility in the hobby; but this can be a mixed blessing. At age 28, without having actually sought the job, Randy Reed finds himself heir to the long tradition of Avalon Hill designers stretching back to Charles Roberts, which is not an easy mantle to assume. Not unlike SPI's Jim Dunnigan, he became involved with a group of players upon which he has continued to draw for Avalon Hill's playtesting. In Reed's case, the organization was Interest Group Baltimore.

Reed's first direct involvement with Avalon Hill was answering questions posed by buyers. It was because of this and other efforts that Tom Shaw finally twisted his arm (after several refusals) to become head of R&D for Avalon Hill. In fact, he thereby became head designer.

Reed agrees with the Avalon Hill idea of a limited number of games per year, but not for the traditional reason that more than a few games per year would glut the market and/or exhaust the available topics or titles. Rather, Reed is concerned with the amount of time needed to do a game properly. As anyone who sees *Tobruk* will agree, a great deal of work went into the game.

One of his concerns is that whenever a game differs from certain conventions, it is criticized, largely because people will not accept the changed conventions. He has certain ideas on game development and design which would involve altering or abandoning many of these conventions. He is also thwarted by the need to try to cover every loophole, and has expressed a preference for general rules, clearly conveying intent, which can then be worked out, in the event of any problems, in light of the expressed intent. Whether this is possible in light of the nit-picking "blood lust" to win which seizes so many players remains to be seen. Most interesting is Reed's desire to explore, even more closely than in *Tobruk,* the interface between miniatures and board games.

Randy Reed, given instant fame, has also acquired the instant burden of not only living up to Avalon Hill's reputation, but to its goals. How accurately he reads which of his ideas the members of the hobby will accept will determine how successful he will be. All of this should make the next few years at Avalon Hill interesting.

John Hill

The list of designers known by name remains small. A figure with significant renown, primarily because of one game *(Bar Lev),* is John Hill. Until recently, he *was* Conflict Games Company.

John Hill has a most conventional background, having neither dropped out of high school nor college, and having attained his degree in American military history. He has had one advantage over other prospective game designers: he owns his own hobby shop. This not only provides him with a source of income for the more mundane things, such as food and clothing, but has also made the difference when it came to marketing his games. Having dealt with salesmen in the field, he was able to make important contacts rapidly in order to promote the sales of his games.

The result was that John Hill's games got rapid distribution in the Mid-West. Since only Avalon Hill and SPI were in the retail field, this served to reduce competition from other designers who dealt only by direct mail.

Despite this initial success, he recognized the inherent advantage SPI has in having a magazine for instant advertisement of new products. Accordingly, he worked out an arrangement with Simulations Design Corporation which resulted in both his company and SDC using the same symbol. At the time, it seemed like a good idea, but with the problems that *Conflict* magazine has gone through, it no longer appears so clever.

This association was augmented by his doing games for SDC, to be published in their magazine. One of SDC's more popular games, *Battle for Hue,* is an example of this collaboration. However, there never was any legal connection between the two companies. Nevertheless, when it became apparent that *Conflict* magazine was in trouble, John Hill's Conflict Game Company was called to account by subscribers. Certainly this was a logical conclusion to reach in light of the deliberate attempt to tie the two together in the minds of the public. It just did not happen to work out as well as was hoped.

Although he has no practical military experience, his educational background has led him to emphasize modern games, though there are notions of Napoleonics in the back of his mind, and his *Verdun* game show that he has an ability to design a pre-World War II game (which so many people apparently cannot do). This educational background also led to his emphasis on artillery and the "first fire" concept which seems to have become a John Hill Hallmark.

Perhaps an equally important contribution is his ability to recognize his own strengths and weaknesses. He feels that his major contribution is the development of a game system. Certainly, he has not used the same game system twice. Better than most designers, he can develop a game system which brings out the unique concepts of the historical situation in question. The key, then, is development. In this he is fortunate in having two loyal friends who are willing to labor in the vineyards to develop the game, as well as having a reliable pool of playtesters. For he recognizes, as do other major companies, that a designer cannot effectively develop and playtest his own game; he is too close to it to see the flaws and problems. Most often the real cause of disaster in game design lies in the failure of the developer or playtester to identify and/or correct problems in the game. Nonetheless, he remains the primary exponent of the chatty, sometimes tongue-in-cheek rule system. Like Randy Reed, he is bothered by the idea that every loophole must be plugged and that the rules must read like a statute to prevent misinterpretation. Rather, he prefers to consider wargamers as gentlemen who can, when presented with an apparent hole in the rules, work out a solution based on reason. This has resulted in some of his more notorious instructions, such as to "take any reasonable supply rule."

The collapse of *Conflict* magazine, coupled with the increasing growth of Conflict Game Company, forced Hill to re-evaluate his position. He has a successful hobby shop. At the time he made his decision on the game company, it was bidding to become a second livelihood. He obviously could not devote adequate time to both, with the prospect that neither might flourish. So, when Game Designer's Workshop approached him on the possibility of selling the company to them, he was interested and, of course, ultimately did so.

In the short run, this was a good thing all around. It allows GDW to rework some of his graphics, combining their strongest point with his weakest. It also allows him to design games at his own pace and not worry about the day-to-day mechanics of running a company (which can be a full-time operation in itself). On the other hand, it does not ensure him a ready market for his games. A game such as *Verdun* would appear to have little potential because it was such a bloodbath and involved so little movement. However, taking a fragment of the battle, he was able to add the element of mobility. He now has the problem of convincing someone to publish a game like that. It may not be a great problem, but it is one which never existed for him before. So, as with so many things, the sale of Conflict Games Company can be a mixed blessing.

For the future, he plans to design games as he has in the past, and at about the same pace. This time he will have to find a publisher for the games. That should not be too hard in light of the reputation he has obtained. Selling the company is of little importance if the "cult of personality" is a valid basis for predicting the sale of games. As one of the few designers known by name in the hobby, his future success may well determine how valid the "name" designer theory really is.

John Young

This native New Yorker was one of the original members of SPI, although he is no longer active in the field of wargames. With a background in accounting, Young served as Treasurer/Accountant for the fledgling corporation, designing games in his spare time. The number and excellence of the games designed by Young pay tribute to his fertile imagination: *La Grande Armee, Musket & Pike, Dreadnought, Borodino, Seelöwe,* and others. Young's strong points were his perception of the gaming problems surrounding the pre-20th Century era and his ability to design systems for these periods, as can be attested by the success of his Napoleonic and pre-Napoleonic games.

Irad "Terry" Hardy

Hardy, a 33-year old Harvard graduate with a degree in American history, is presently the head of Research and Development at SPI. Coming to SPI in late 1971 as Operations/Business Manager, Hardy cut his designer's teeth on the interesting, if little-lamented, *Scrimmage.* However, Hardy's ability to assimilate and employ the minutae of all eras of history (his particular fortes are the American Civil War and modern tactical combat) quickly led to such block-buster games as *Firefight, Chickamauga, Patrol, Mech War '77/Panzer '44, Punic Wars,* the immense *War in Europe,* and the forthcoming *War Between the States.* Hardy's credentials and the influence of his games label him as one of the leading designers in the industry.

Frank Davis

Known at SPI as Mr. Meticulous, Davis' forte is in the area of rules. Davis, now 29, started work at SPI as a shipping clerk. An avid reader and a devotee of accuracy, Davis' first effort was as developer for *Operation Olympic.* His first complete design was the critically acclaimed *Frederick the Great,* which revealed his penchant for meticulous, intensive design-work based on loophole-free rules. Presently at work on the massive *Wellington's Victory,* Davis first steeps himself in as much historical material as he can, so as to obtain a complete feel for the period he is working on; it is only then that he begins working.

Richard Berg

Originally known in the hobby as the widest-read game critic, it is only recently that Berg began designing actual games. A lawyer by profession, Berg's first design was *Hooker and Lee.* However, it was with his second design that Berg rose to prominence in the industry; *Terrible Swift Sword,* an immense tactical-level recreation of the battle of Gettysburg, proved an instant hit. Using his background as a lawyer, Berg uses a problem-solving approach to game design, and he often specializes in devising systems to fit a particular era. His most recent work includes *Conquistador, The Conquerors,* and the soon-to-be released *Lord of the Rings.*

David Isby

Isby, a 24 year-old Columbia University graduate, has been SPI's Director of Research for the past several years. He fell upon his job at SPI when he ran into Jim Dunnigan while the latter was working as a night-watchman for General Electric. Aside from providing general and specific information for a wide variety of games—especially for air games—Isby has designed *Mukden, Marengo,* and *Crusader.* His initial design, *Soldiers*, is still considered one of the finest WWI games available, and his soon to be released *Air War '78* attests to his excellent grasp of that field.

Al Nofi

Nofi, a 33 year-old teacher, and now one of the guiding lights of the new Morningside Project, is the former head of Research at SPI, as well as being one of the Ur-staff members. Considered a walking encyclopedia by a number of his co-designers, Nofi's area of expertise is in pre-20th Century warfare. His designs include *Centurion, Renaissance of Infantry* for SPI, as well as *Vicksburg, Wellington in the Peninsula,* and *The Great War* for Morningside.

John Prados

Prados, a native-born Puerto Rican, has an extensive list of academic credentials from Columbia University (as do a goodly number of other designers, it seems). The other key design member of Morningside Project (with Al Nofi), Prados first came to the industry's attention with his *Year of the Rat* game for SPI. His *Fall of the Third Reich* for Avalon Hill is still one of the most popular games around, and his *Von Manstein's Battles* was Morningside's first game. His newest release, *Pearl Harbor,* for GDW, again attests to his interest in WWII and later periods.

The GDW Team

Perhaps the most "together" military team since the Three Musketeers, the team of resident designers for the estimable Game Designers' Workshop includes Marc Miller, Frank Chadwick, Rick Banner, and Loren Weissman. The first three had worked together at Illinois State University under a grant, developing conflict simulations for various projects. When the money ran out and the project fell apart, the three decided to form their own company and produce their own games. Their intent, originally, was to produce a giant WWII game—now known as *Europa*—and this was primarily the work of Rick Banner, who also does most of the art work for GDW. It was Banner who masterminded the gigantic *Drang Nach Osten* and its companion, *Unentschieden,* which led to the latter games of *Narvik* and the Sealion portion of *Their Finest Hour.* Marc Miller also did a great deal of work on *Their Finest Hour,* in addition to his designs for *Triplanetary, Chaco, Coral Sea,* and *The Russo-Japanese War.* Designer Frank Chadwick, considered the "systems man" of the group, worked with Banner on DNO etc., while also designing *Torgau, Crimea,* and *Avalanche.* Weissman, whose area of experience is in the ancients period, did the design work for *Eagles!.* As can readily be seen, the GDW designers are quite eclectic in their tastes and do not seem to be limited to any one particular area.

Steve Peek

A designer who virtually backed into the job, Peek ran the small but prolific Battleline Company single-handedly for over three years. Starting with an idea for doing a Civil War game—*Seven Days Battles*—Peek slowly built his company into one of the leaders of the Third World of gaming. Concentrating on unusual approaches to games, Peek had instant success with *Wooden Ships & Iron Men* (actually designed by his partner, Craig Taylor), and *Shenandoah,* a highly unusual Civil War game. Although his area of preference is the Civil War, his work in such areas as naval warfare *(Submarine)* show that he has the flexibility to become a top designer.

THE OUTGROWTHS

The 'Zines

Over the years since the introduction of the Avalon Hill *General* in 1964, there has been a remarkable proliferation of magazines and newsletters devoted to the wargaming hobby, ranging from the amateurish, spirit-duplicated club newsletters, through high quality, professionally printed, sophisticated amateur publications such as *Panzerfaust, Spartan International, International Wargamers,* and *The General* itself, and ending with the professional *S&T* and its spin-off, *MOVES*. There has been a steady turnover in the ranks of these publications as older ones die out for lack of interest and are replaced by newer ones. Several 'zines supposedly published by clubs are actually labors of love of one or two individuals, such as AHIKS' *Kommandeur.*

A recent trend in wargaming 'zines is toward magazines that might be termed "thematic" publications that focus on a single period, or even a single year, or single game, such as *The Barbarossa Bulletin,* dealing with the game of the same name; *Fusilier,* dealing with the Napoleonic period; and *Europe '44,* dealing with events in Europe in 1944. In addition, and to some extent partially outside the "wargaming" hobby are the *Diplomacy* 'zines, such as *Graustark, Erewhon,* and *Atlantis;* and the *Origins* 'zines, devoted to playing out entire games through the pages of the magazines in question.

In addition to the increase in the number of 'zines, the physical and editorial quality of the publications has also improved. Thus, *The General* for a considerable time looked more like a newspaper than a magazine, and was printed on cheap paper with very few illustrations, no real format, and no color. Issue 1 of Volume 3, out in May 1966, introduced a more magazine-like cover, with a colored border and an illustration, but had even fewer interior illustrations. With Volume 4, Number 4, out in November 1967, *The General* went over to coated stock, while still keeping interior illustrations to a minimum. This has been the format ever since.

Editorially, a number of changes have occurred. The first issues of *The General* had regional editors, each supposedly recruiting articles from his area. The early articles were on such topics as "Waterloo—A Defensive Strategy," "Le Mans Chance Table," and "Hints for Stalingrad." Most of these articles ran no more than 1,200 words, and usually much less than that. The more recent issues have somewhat longer articles, often pompously titled, "Simulating the Art of War" or, more obscurely, such as "Operation Morgenstern," but also many of more direct wargaming application, such as "1914 Strategy and Tactics" and "Stalingrad: The Middle Game." A regular feature of *The General* has been the "opponents-wanted" page and the contest, in which a wargame problem is presented to the readers, and they are asked to propose solutions using standard PBM (Play-by-Mail) techniques. Almost every issue also has a question and answer column, where rule hang-ups are examined and clarified. In addition, each issue now features a play-through of a game, with comments by the players as well as a neutral judge.

The oldest of the club 'zines, *International Wargamer* and *Spartan International,* began to fall by the wayside for various reasons. The collapse of IFW and Sparta as viable organizations took their magazines out of circulation, though the latter magazine has since reappeared. *Panzerfaust* survived the transfer from Don Greenwood, its founder, to Don Lowry who, in turn, carried it through a move from his Illinois home to Maine. But then, for various reasons, it ceased publication for a year. When it resurfaced, along with Don Lowry, it was in California. The name *Panzerfaust* was shoved into the background in favor of *Campaign.* Aside from the name and the lineal trace, the relationship of the present magazine to its predecessor is tenuous. Even so, it now publishes regularly and has a strong, if small, following.

One of the longest-published club 'zines in existence is AHIKS' *Kommandeur.* Well into its twelfth year of steady informational service, *The K,* as it is fondly referred to, is the organ of its parent, AHIKS, the international Play-by-Mail club. While it could be pointed out that *The K,* under a variety of editors, epitomizes the best in club magazines, it is of more interest to note the place in wargaming history that *The K* and AHIKS might have had. It seems that when Chris Wagner was searching for someone to revive his stuttering *S&T* in the late '60's, one of its more prolific contributors, Henry Bodenstedt—one of the founders of AHIKS—considered incorporating *S&T* into *The K.* The idea never grew past the thinking stage, and Dunnigan eventually took over the helm at *S&T*. *The K* remained a competent, informative newsletter; it would be interesting to speculate what would have happened had it taken the plunge.

Perhaps the most successful of the recent 'zines is the Texas-based *Jagdpanther* (now *Battlefield),* brain-child of Steve Cole and Allen Eldridge. *Jagdpanther* started life several years ago as any one of a number of hastily-printed magazines that featured variants and other such articles on already existing games. But Cole was smart enough to see that the one item that had "made" *S&T*—the inclusion of a game with each issue—should be good enough to make his publication more than just a fan 'zine. So *Jagdpanther* started to publish games with each issue beginning around 1973–4. By 1975, the magazine had progressed to the point where it was printed in glossy format with semi-professional touches, and the games enclosed were no longer pieces of stray paper and random bits of cardboard. By the end of 1976, *JP* had achieved a normal publication rate, its games were professionally printed with standard components, and the name of the magazine was changed to *Battlefield.* While the general editorial intent of *BF* still centered around variants (few articles ran more than half a page), the success of the magazine attested to the popularity of the formula.

It is hard to conceive of the gaming public absorbing more magazines than there are available in 1977. The emergence of *Fire & Movement,* Rodger MacGowan's professionally produced publication filled with such comment and criticism by some of the top names in the field—Mark Saha, Randy Reed, Jim Dunnigan, etc.—was greeted enthusiastically. Yet the enthusiasm was reserved, for few people within the industry gave *F&M* more than a fighting chance. Few magazines have achieved more than a minimal subscription rate; many have gone by the wayside. As the hobby continues to expand, it will be interesting to see whether that expansion includes a broadened and critical interest in the hobby itself, for only with that type of expansion will the various magazines be able to survive.

The Clubs

The rise and development of the clubs is an event worth recounting in itself. This is due, in part, to the fact that the clubs are both the most visible and most organized part of the hobby.

The nationwide club was a virtual impossibility prior to the advent of *The General.* This is, as has been noted elsewhere, due to the fact that there was no real means of knowing who was interested in the hobby short of some newsletter. To be sure, there were local clubs, particularly at colleges and universities, but beyond that level no formal organization existed.

It is interesting to note that clubs grew and developed in a manner not unlike a child. Thus, when speaking of the immaturity of the early clubs, the analogy is particularly pointed if viewed over the long haul.

It if generally conceded that the first national wargame club was S.P.E.C.T.R.E., obviously named after the then-popular Ian Fleming organization fought by James Bond. In this case the initials meant *Special Efficacy for Counter-Espionage, Revenge, and Extermination.* S.P.E.C.T.R.E. was formed in the spring of 1965, a year after the first issue of *The General.* Before that there were local clubs, early examples in the play-by-mail field being the *Avalon Hill Wargamers PBM League of Huntington, Indiana* and the *Tactics and Strategy Club of California.* But somehow, the idea of S.P.E.C.T.R.E caught on and it clearly outdistanced the smaller clubs. Donald Greenwood, who had been highly visible among the hobbyists, both as publisher of *Panzerfaust* magazine and as a player himself, carved out his own niche at this time. He noted that the filling of the "ranks" (all clubs seem to have a need for ranks, the more martial the better) was on a first come, first served basis. S.P.E.C.T.R.E command was loosely based on the game *Tactics II* in that there were army group commanders, army commanders, corps commanders, and so on. Greenwood got in early enough to be an army commander. The looseness of the selection of officers reflects the general nature of the early clubs. S.P.E.C.T.R.E. charged no dues and offered no services to the members. It was simply the "ego trip" of being in a club which claimed, rightly or wrongly, to have the best players going.

As fast as S.P.E.C.T.R.E. announced its importance, rivals organized to knock it down. This opened the major phase of the early history of the clubs: "world conquest." The technique was one step better than the classic announcement of an ability to "lick any man in the house." A challenge would be put out that a given club "owned" a certain state unless someone defeated the club and proved the contrary. Moreover, when one club seemed to be of importance, the rival clubs would seek members on the grounds that the "menace" of the other club had to be stopped. It had all the elements of a Moslem *jihad.* As fast as S.P.E.C.T.R.E. was organized, a club called *World Conquests, Inc.* was calling for a holy war to stop "the menace" of S.P.E.C.T.R.E.

In mid-1965, there appeared to be two directions that the clubs were taking—those oriented toward World War II, which took such names as *SS Panzer Corps,* and those oriented toward the American Civil War, which called themselves such names as *Confederate Army of Central Pennsylvania.* There were, too, a few which used acronyms, such as S.P.E.C.T.R.E. and DUSK (Der Uberlegen Siegreich Kommando). It was in this period that the clubs can be considered as embarking on their most immature phase. Greenwood aptly refers to clubs of this period as "warring" clubs, to distinguish them from the present "wargaming" clubs, which he considers to be more hobby oriented.

Some efforts were made to add maturity to the hobby, notably the *Avalon Hill International Kriegspiel Society* (AHIKS) which had a minimum age requirement of 25. Unfortunately, they discovered that youth has no monopoly on immaturity.

There is some reason to doubt that many games were ever played to completion by these clubs. Unblemished records were maintained over a large number of games by the simple expedient of abandoning a game when the tide ran against you, thereby being able to say that you had not lost. Few games had only one side claiming victory during the period 1965–1968. The pages of *The General* provide a microcosmic view of it all. Appearing bimonthly, only the major twists and turns were apparent and they became laughable as one club would declare that it controlled an area, a second that it controlled the same area by virtue of having beaten the first club, and a third that it controlled the area, was neutral and had not been beaten by anyone.

Probably the most disquieting aspect of it all was the general drift toward German club names and, most specifically, Nazi-oriented names. This tended to be complemented by a general attitude in keeping with the name. Thus, while the early clubs such as DUSK soon lost the original sense of their name and while a name like *SS Panzer Corps* is fairly innocuous, by 1967 one saw *Wehrmacht III, Gauleiter of San Francisco, 4th Reich* (as well as *3rd* and *5th* at various times), *1st SS Der Totenkopf SS, Kampfgruppe Pieper,* and, still later, *Liebstandarte SS, Panzer Lehr, 1st SS Raumsturmgeschwanter der Totenkopf SS, 9th SS Volksgrenadier, Fuhrer Escuadrille, The Gestapo, Deutschland Uber Alles,* and others of the same ilk. One never got as far as *Konzentrationlager Auschwitz,* but that may have been for lack of time. This is the aspect of the hobby that both Charles Roberts and Thomas Shaw, to name only a very few, found repulsive. Certainly it did little to advance the reputation of the hobby.

The constant nonsense of "conquest of states" began to pall as the members got older. Two organizations began to attempt to sort the matter out. The older would appear to be one originally known as *United States Continental Army Command* (USCAC) which later changed its name to *International Federation of Wargamers (IFW)* and the other, approximately half a year later, being *Spartan*, now *Spartan International*. Both were interested in bringing a little order to the chaos. It is difficult to assess when USCAC determined that this would be a main feature of its efforts but it would seem a reasonable assumption that it did so after *Sparta* since USCAC/IFW has always been more loosely run. *Sparta* was the creation of Russell Powell and, until recently, had been almost a personal fiefdom. It was founded on the idea that the hobby needed a better image and could only gain it if the clubs pulled the hobby up by its bootstraps. IFW, on the other hand, did not appear to have had the same relentless sense of mission behind it. The major move these clubs made, which allowed them to rise above the pack, was to eliminate the warring element and return to the hobby. Secondly, they offered services to the members—a newsletter (in fact a number of newsletters, aimed at different phases of the club) and listings of opponents wanted as well as organized tournaments and conventions.

These two clubs were and remained a study in contrast. As mentioned before, *Spartan International* was the creation of Russell Powell. It was primarily through his fixed idea of where the hobby should be and his relentless insistence on reaching that goal that *Spartan International* grew. Obviously, such a rigid approach brooked no major dissention and this, in part, explains the fact that the IFW is the larger of the two. Whereas *Spartan International* went out of its way to eliminate aspects of the hobby which it believed detrimental (notably the idea of "conquest"), the IFW was more loosely controlled. There were personal attacks in the IFW publications recalling the "glory days" or vitriol. *Spartan International* would not consider printing such remarks.

Powell, as the sole head of *Spartan International*, presented a more visible target than any individual in the IFW. The IFW not only changed leadership frequently, but even moved its center of operations from the east coast to the midwest.

Perhaps as a sign of greater maturity, the clubs began to fall on hard times. IFW was always looser than *Sparta* in its organization and seems to have disintegrated by centrifugal force. *Sparta* disappeared under more suspicious circumstances. Powell stepped down as head (some allude to a palace coup) and his successor could not maintain control so that when a part of the treasury went east, the organization fell apart. Attempts to revive the club have not succeeded, and there is really no national wargaming club of the impact that these old clubs had.

There are still clubs devoted to wargames, however. AHIKS, originally the *Avalon Hill International Kriegspiel Society* was founded to provide an organized means for playing wargames by mail, and it does provide a nationwide continuity, which is the best to be expected of a club. Its early connection with Avalon Hill (in name only) has been entirely abandoned, and AHIKS now provides PBM services for all wargames. It has its own 'zine, the *Kommandeur*, and affords a foolproof and simple system for mailing moves. Players are ranked, but the competitive aspect of gaming is minimized. AHIKS seems content with its low profile rather than making deliberate attempts to dominate the hobby, as *Sparta* did. Some of the most respected names in the hobby are members and the membership is truly international and growing.

And, of course, there are other, more local clubs, the most prominent of which—thanks to their symbiotic relationship with Avalon Hill—is *Interest Group Baltimore*, the group formed originally to playtest AH games.

The warring clubs had a certain attraction in their bald self-confidence, but since they have been discredited, it would appear unlikely that the hobby is ready for a truly international organization at this point. This is not to say that it will never have such an organization, but the present temper of the times is that wargamers are individualists. Since they cannot agree on which is the best game, it cannot be expected that they will agree on letting someone else tell them what is the best wargame to use in resolving competitions. Clubs, at this point, are in a state of limbo in the wargaming evolutionary line.

The Conventions

In addition to taking over the idea of fanzines from science fiction fandom, wargaming also adopted the convention, or "con." The idea for a convention seems to have originated almost as soon as the first issue of *The General* hit the stands. To date, a number of local wargaming cons have been held, and in 1975, the first National Convention took place.

Most of the cons have thus far been one club affairs, during the era when the clubs were viable. There was the *Mil-Con* of the *Military Strategy Confederation of New York,* or *Sparta's* annual *Eastcon* at the battleship *Massachusetts.* The longest continuous con has been the annual affair held at Lake Geneva. Originally "an IFW" con, it continued after the club had folded and is now known as *GenCon.* Under the aegis of TSR Game Company, it is the most popular of the local cons. Its flavor, however, is a mixture of boardgames and miniatures, as well as *Dungeons & Dragons.*

It took seventeen years for the first serious attempt at a national convention: *Origins I.* This convention, sponsored by Avalon Hill, was the first to be supported by *all* the major wargame publishers with the result that it got widespread publicity. Held in 1975, It immediately excited interest in the hobby in a rotating convention, held in various locations around the country. *Origins II* was held with even greater success in Baltimore the following year. *Origins '77* is slated to be held in New York City.

At the same time, the first serious attempt was made to give awards for achievement during the previous year in terms of best magazine, best game, and so forth. A convention is an obvious vehicle for doing this, and a national convention that is not under the thumb of one company or another gives greater credibility to such awards. Awards, however, are still in their infancy, and it will take a major publicity effort to alert gamers to their existence. In the meantime, what awards there are remain rather provincial and of dubious worth.

Wargames Beyond the Seas

Approaching the end of its second decade as a genuine commercial venture, board wargames are still not as popular outside the United States as they are inside. They also have far greater popularity among English speaking peoples than among others. By and large, outside the United States, those who play wargames use miniatures rather than maps and counters.

Although wargaming's popularity outside the United States would appear to be greatest in Canada, activity there has been very much in the orbit of activity in the United States. There is no *major* Canadian wargaming magazine, either of the *S&T* type or like England's late, lamented *Albion.* On the other hand, there are game designers in Canada, including Gamma Two, with its unique approach to Napoleonic games. Moreover, one of the more successful wargames in Britain was designed by a Canadian *(Decline and Fall).* So wargaming is alive, if not exactly thriving, north of the border.

In Britain, the situation is perhaps not as good. For a number of years, Donald Turnbull, through his magazine *Albion,* was the focus of British wargaming. Unfortunately for British wargaming, as well as for those who enjoy an interesting wargame magazine, *Albion* became less a labor of love and more of a labor; so, with issue 50, *Albion* ceased publication. Nothing has arisen to quite replace it, and there is nothing in Britain on the physical level of *S&T, MOVES, The General,* or even *Conflict* or *Panzerfaust/Campaign.* On the other hand, wargaming in Britain is not totally without focus. Malcolm Watson acquired the SPI interests in Britain some time ago and has formed SPI/UK.

More importantly, he handles most of the good wargames from the U.S., as well as some British games, making his operation a focus for Britain since he is the source of what is new and current in the hobby in terms of games.

As far as wargaming is concerned, AHIKS served as a first wedge of boardgaming into the heretofore miniature hobby. AHIKS was the only organization which promoted national events. In 1973, the National Games Club was formed, but it was not solely a boardgame organization, and in that respect it is typical of so many areas in Britain insofar as boardgames are concerned: they are represented, but they are not the sole focus. In Don Turnbull's view, *Diplomacy* has thrived at the expense of boardgaming, picking up those who find the games too difficult. It is more difficult to get a handle on the British scene, because no company or magazine occupies a position comparable to SPI/*S&T* or Avalon Hill/*The General.* Among the more active wargamers in Britain, there is the feeling that the British scene is quite viable but is still awaiting the catalyst that will make it as visible and apparently cohesive as that of the U.S. (if one can say that wargaming in the U.S. is cohesive).

It may well be that Malcolm Watson, in his position as the principal supplier of games, will permit his magazine, *Pheonix* (sic), to rise to the top since in that 'zine readers will get first-hand information, rather than reaction to what someone else has done. The fact that no one has arisen in Britain comparable to Avalon Hill or SPI or even some of the smaller companies in the U.S., with the purpose of putting out boardgames of the type familiar to U.S. wargamers, is one of the major curiosities since U.S. boardgames do sell well in Britain despite a markup. For example, an $8.00 game in the U.S. will sell for the U.S. equivalent of $9.10 which, in the present state of the British economy, represents an even greater outlay then the simple price differential indicates.

In the wake of *Albion's* demise, European wargamers now have *Europa.* Published in English in Switzerland by Walter Luc Haas, *Europa* must stand unique in the hobby. First, it is a substantial size and, being mimeographed, represents a great deal of work in cutting stencils, running them off, and assembling the finished product. The reproduction is fairly good. The most interesting aspect of it is the fact that it is a giant "letters to the editor" column. Editing seems to consist mostly of grouping, in that Mr. Haas does not contribute much copy himself, reserving his activities more for commentary on others' remarks (which is not to say that a reader does not know where he stands on various issues). A given article might be a variant article, a game review or a play technique article. Mr. Haas makes use of the interesting device of juxtaposing articles on the same topic which espouse different principles. The articles are, for the most part, written by Americans, though there are a few by people from non-English-speaking countries. Considering its point of origin, the language is surprisingly good idiomatic English.

Among non-English-speaking Europeans, boardgaming is marginal. However, both Avalon Hill and SPI are engaged in working out arrangements for foreign language versions (initially German). Whether, in light of the British situation, there is a market for wargames outside of this country, especially one which warrants the cost of translation, is a gamble which both companies are willing to take.

Of the major English-speaking countries, only Australia has made a venture into the boardgame field in the sense that it is known here, with Jedco's Russian and North African campaign games. Both use hex grid maps and fit into the "mainstream" of American wargame design. On the other hand, no Australian wargame magazine has arisen which is on a par with *Albion,* much less the more professionally printed publications of the U.S. Since one of the cliches common about Australia is that it is much like the United States (or, rather, as the U.S. was), it may be that Australia will ultimately become the biggest non-American market for wargames.

Wargames are still in their infancy outside the United States. It may be that each country is waiting for its version of Avalon Hill or SPI before the hobby really catches on.

MAJOR LANDMARKS

Wargaming can be measured in terms of certain milestones, the vast majority of which relate to innovations in game design. Sometimes these innovations are forgotten and reinvented with the reinvention gaining better recognition than the initial development. In other cases, innovative ideas are not used again and become blind alleys. What follows is a list of some of the major landmarks in the hobby and, in the case of games, the landmarks they contained.

1953 Charles Roberts designs *Tactics* and markets it commercially. Being the first game, all of it is innovative, but among the more significant elements are:
- zones of control
- square grid

1958 The Avalon Hill Company is formed by Roberts.
Gettysburg (AH)
- first historically based wargame
- unit facing

1961 *Chancellorsville* (AH)
- hexagonal grid system
- secondary zones of control

U-Boat (AH)
- semi-hidden movement
- tactical naval warfare

1962 *Bismarck* (AH)
- fully hidden naval movement
- strategic and tactical levels

1964 Avalon Hill publishes *The General,* first commercial wargame magazine
Afrika Korps (AH)
- overrun rule
- supply units

Midway (AH)
- separate air units and air rules

1965 *Guadalcanal* (AH)
- step reduction record
- defensive artillery
- ranged weapons fire

1966 Christopher Wagner publishes *Strategy & Tactics*

1967 *1914* (AH)
- semi-active zones of control
- inverted and dummy counters
- multiple combat results tables
- reduced strength counters
- refitting of units

Vietnam (GS)
- area movement

Confrontation (GS)
- multi-player game
- point/linear movement
- production of forces

1968 *Battle of Britain* (GS)
- first air warfare system
- "brick" grid system
- skill level for veterans

Trafalgar (RC)
- first multi-scenario game
- first tactical sail warfare game
- point control of movement in lieu of grid

1969 *Hannibal* (LR)
strategic ancient warfare
land combat in same hex
Blitzkrieg Module System (PP)
modular rules
multi-phase movement and combat
Deployment (PP)
line and column formations
Crete (PP)
first magazine wargame
airborne assault simulation
Flying Fortress (PP)
linking of games for campaign
secret/variable victory conditions

1970 *Kriegspiel* (AH)
matrix combat results table
geomorphic map
Tac 3 (PP)/*PanzerBlitz* (AH)
line of sight/fire
modern company level warfare
fire-before-movement
vehicle silhouette counters
Leipzig (PP)
historically evaluated leaders
Strategy I (PP)
chit randomizer in lieu of die roll
indexed rules
Renaissance of Infantry (PP)
leader units for command control
panic/morale effects
James Dunnigan and Redmond Simonsen incorporate as Simulations Publications, Inc.

1971 *USN* (SPI)
numbered hexes
Lensman (SI)
first science fiction game

1972 *Napoleon at Waterloo* (SPI)
first game designed as introductory game
1812 (SPI)
area and hex versions
Flying Circus (SPI)
air-to-air combat
quasi three dimensional play
American Revolution (SPI)
strength differential combat results table
Quebec 1759 (GT)
"domino" alterable counters

1973 *Battles of Bull Run* (SPI)
simultaneous movement
Sniper! (SPI)
individual-man level
Battle of Hue (SDC)
"first fire"
Fall of Rome (SPI)
first true solitaire game

1974 *StarForce* (SPI)
full three dimensional play
non-linear movement
Bar Lev (CGC)
different scale for different fronts
Tank! (SPI)
counters without any data
Europa Series (GDW)
series of linked games
Prestags Series (SPI)
integrated multiple-era games
Third Reich (AH)
strategic attrition-combat

1976 *Invasion: America* (SPI)
"untried" inverted counters
Dixie (SPI)
expenditure of administrative points to take any action
Starship Trooper (AH)
first simulation of a novel
Terrible Swift Sword (SPI)
ammunition supply and casualty count
"split-fire"

SECTION II:

Game Design and Development:

A Case Study

by Richard H. Berg

Designing a game is very much like writing a book: the designer—as does the author—gathers a great deal of information, marshals it, transforms it into a product, polishes it through many rewrites, and then presents the finished product for consumption and enjoyment by the public. The finished product, in itself, is much like a book, in that the player reads it and uses it for the information it is giving him. The major difference is that the game enables the player to put the information to direct use through the mechanism of the rules and game-parts. And the designer must remember that the game is to be used in this way if he is to be successful in his work. There is no use presenting information that is so obscure and opaque that the player-/reader can't make heads or tails of it. Clarity of thought and presentation and a definite idea and purpose are necessary.

In order to best illustrate how a game is designed (apart from the physical aspects, which are covered later) let us look at an actual game from beginning to end of the design development process. In this way we can trace the progress of game design with reference to a living example.

SPI's *Panzergruppe Guderian* is an excellent subject with which to work. The game, released in mid-1976, proved to be an instant success for its playability, design conception and clarity. The game covers the German advance to Smolensk in the summer of 1941 and the subsequent Russian halting of such advance. It is a game with a clarity of purpose in that it sets its sight on a definite aim and does not get sidetracked along the way with frills and chrome. But the end result of the design is a product of several months of extensive work and rework and good solid research.

The idea for doing *PGG* belongs to wargaming's most prolific and renowned designer, Jim Dunnigan. Most games are designed for a reason, and *PGG* was no exception. Unlike a goodly number of games, it was not published to fill a particular gap in history. Many games *are* so designed: e.g., *Chaco* appeared because no one had ever done anything on South American warfare; *Alexander the Great* was designed simply because there was no game on this most important historical figure. So games can be designed on subjects that have not been covered before. This usually occurs more frequently with "amateur" designers who have pet projects and historical interests. These areas are often somewhat obscure and, for that reason, are rarely covered by the major companies, because obscure subjects generally produce too few sales. Thus, a game such as *Mukden, 1905,* which covers a single battle in the Russo-Japanese War, is published by a private entrepreneur, whereas the idea never even occurred to the larger companies. Fledgling designers should consider this when casting around for a title; there are plenty of uncovered areas of historical interest (e.g., The Taiping Rebellion in China, 1840–54, the Mongols, the Boer War, etc) that have not yet been the subject of a game.

Dunnigan, like a good number of designers, likes to get right to the heart of matters from the outset. As a result, one of the first portions of the game covered was the Combat Results Table. The CRT is the heart of any game; each and every game lives and dies with its CRT as it directs the operations and flow of the game-life. Dunnigan decided on a simple CRT for *PGG.*

Another reason for publishing a particular title is cover an area that has been "gamed" before, but do it in such a way that it throws an entirely new light on the subject. *Afrika Korps,* by Avalon Hill, first appeared in the early '60's and proved an immediate hit. The subject—the desert war in WWII North Africa—was a highly popular one, and the type of warfare it highlighted—fluid movement with few units—appealed to a wide variety of gamers. But *AK* was certainly not the final word. The historical accuracy of the game was questioned immediately and it wasn't long before other designers put in their two cents' worth on the subject. Variants (rules using the original games components but adjusting the game to meet the particular designer's concept of events) appeared almost immediately, followed by several amateur designs, including Loren Sperry's *Rommel* and Chuck Lane's *Desert Fox*. SPI followed with their popular *PanzerArmee Afrika* and the race was on. Each game's designer concentrated on areas he thought should be emphasized. The time and space scales varied, and the game systems were as numerous as the games themselves. But the flow of North African games has yet to stop. In 1975–6 alone there were a host of games, including items by Rand, Balboa, SPI, Avalon Hill, Conflict, Wellington Games (in Sweden); and 1977 will see a giant four-map historian's treatment of the subject on a battalion level. Therefore, another item for a designer to consider is the popularity of the subject and the treatment(s) it has been given before. Perhaps the designer feels he has something new to say, something exciting, something the players want and will play. To simply do a subject over again because it is your favorite is not enough; make sure there is something to say.

With *Panzergruppe Guderian* the reasons for design were somewhat unusual, and a bit mercenary. *Strategy & Tactics'* (SPI's history *cum* game magazine) recent games had suffered severely in the ratings, partially because gamers resented the smaller, "folio" size and partially because of the subject matter. Jim Dunnigan felt that a change was necessary and he quickly looked for a game on an area of interest that would produce sure-fire results. The choice was almost obvious: Russian front games have always been popular (to say the least) and one of Dunnigan's fortes is that particular area of history. The early portion of the campaign had received little attention on an operational level, and the Smolensk offensive presented itself as an excellent topic for simulation. Thus the designer's familiarity with gamers' desires, plus a working knowledge of the subject, combined to produce the needed game title. As with most SPI games, the title was placed in the reader feedback (SPI has that advantage over the amateur designer; it can see whether its game will sell *before* it is designed!), and the results proved Dunnigan's thesis correct: *Panzergruppe Guderian* was a hit in the feedback. All that was now needed was a game.

The SPI system of game design is quite simple; unfortunately it is a luxury that few amateur designers can afford, simply because they have neither the resources nor the manpower to handle their product in such a fashion. However, it is worth considering simply for the light it throws on the design process. At SPI a two-man team handles a game: the designer does the research and initial conception; the developer then takes it from the designer and polishes the product, consulting with the designer as he proceeds to a conclusion. This system has several major advantages. First, it puts two people's experience into a game. This usually creates an atmosphere where ideas and concepts are nurtured as the designer and developer compare notes. Unfortunately, it sometimes means conflict. If the designer sees the game differently from the way in which the developer is proceeding, then trouble can arise. Thus, the dual production system requires discipline, professionalism, and the ability to take and receive constructive criticism. Secondly, the dual system enables a designer/developer to handle more than one project at a time, which is important if you're in the business to make money.

In any case, each designer has his own particular point at which he will give a game to a developer. Much of this interrelation depends on how much control the designer is willing to relinquish and how much trust he places in his developer. Again, discipline and professionalism are needed.

The first step to any design is the research, and for *PGG* Dunnigan's background proved quite handy. Knowing exactly where to look for the material he needed, he spent an inordinately few hours gathering the necessary material. Most amateur designers will find that a great deal of reading and notetaking must be done before sufficient material is available from which to start. First a general appreciation of the subject must be attained, then the minute facts—

the hardware—must be dug out. In the latter case for *PGG* that meant procuring the Order of Battle. It is tough to proceed with your design without knowing who was present historically, so this type of information is usually extracted as early as possible.

Using several recently published sources as well as some standard volumes on German OB, Dunnigan was quickly able to determine the designations of the various Army-size units that were in the area of the Smolensk offensive. German information on divisional and regimental size units was also readily available, but similar information on Russian units was scattered and diffuse at best. The numbers were available, but the exact whereabouts of the units at any given time was in question.

This problem was further exacerbated by the fact that the author of the S&T article that accompanied the game in the issue was coming up with different information from that which Dunnigan was producing. And this is a prevalent problem with designing historical simulations: it is unusual to obtain solid, exact information on any given subject. There will invariably be confusion and contradiction. It then becomes the designer's problem to use this conflicting information. Does he pick and choose between sources, or does he go with one interpretation for the sake of uniformity? This is an important decision, and the designer's familiarity with his subject will often balance the scales in one direction or another. Basically, if you are not particularly familiar with your subject, you should pick a source and stick with it, unless another source convincingly changes your mind about a certain fact. If you are well-versed in the area, you can afford to be selective, knowing instinctively which information is more trustworthy. In such a manner was the OB for *PGG* worked out. The next step was the game-map.

Drawing a game-map is relatively easy in terms of physical work (see the section on physical components). Finding a map source is something else. Many geographical areas are charted, but the reliability of some of these maps can be questioned. Chief among these would be any area in the Soviet Union. The Russians are reluctant to reveal any information of this sort, and finding a reliable map of the Soviet Union in acceptable detail from any time after the Revolution can be a chore. It is not unusual to find two maps with the same city located in two areas that are 50 miles apart or rivers running completely different courses. Maps on Napoleonic battles or the American Civil War are much easier to come by, but even they can be troublesome. (The West Point Atlas, which is used by many amateur designers for the Civil War is notoriously inaccurate on that subject!) Designers must be careful as to their terrain source; and the more tactical the game, the more definitive they will want the terrain—and the source—to be.

With *Panzergruppe Guderian* the first choice was, unfortunately, the wrong one. Dunnigan used one of the old SPI Russian front games as his basic source. Unfortunately, the source for *that* game *(Destruction of Army Group Center)* was none too good and the result, somewhat realigned and blown up, was a fairly inaccurate map of the Smolensk region (a fact that was not realized at the time). This is a hazard that all designers must encounter, a hazard that is sometimes difficult to spot. Usually—as was the initial case here—the end result will be a warped result, a result that is occurring continually and in defiance of all history and knowledge. This something that will come out only after testing, a task that can be avoided by making sure that you have good sources to begin with. For the amateur, there may be no way of rating a source, so he will have to trust a) his local librarian or mapseller, and b) his instincts. Sometimes the latter are better, but usually the former can be of immense assistance. Dunnigan now had a map, (albeit an inaccurate one) and an OB. He now needed a system—and a game.

The general outline of the actual campaign showed the usual German blitzkreig–like breakthrough against disorganized and widely variable Russian divisions. However, initial German successes quickly disappeared as pockets of resistance formed—usually around major cities—and a constant stream of Russian reinforcements managed to form a line east of Smolensk to halt the German advance, even to the extent of launching a counter-attack late in the summer. Such a situation was not, in general, unfamiliar to wargamers. However, certain problems presented themselves, and from these emerged the major design concepts inherent in the game.

The idea was to keep combat simple, allowing the flow of the game to proceed unimpeded. Fancy and intricate combat systems can slow playability to a crawl: *Tobruk* and *Highway to the Reich* are prime examples of this type of CRT. In both of the above games, the designer was intending to produce a definitive statement about the respective events, and to that extent both succeeded. But playability suffered, as it will when a designer tries to get more and more information of an admittedly complex nature into the game. The designer must therefore decide what the thrust of his game will be. He also must, of course, consider the era he is simulating.

CRT's in and of themselves are not difficult to design. The basic Avalon Hill/NAW (SPI) AE-DE-Exchange CRT can serve as a starting point for any novice designer. It incorporates many basic precepts about combat ratios and results that still work today. They are, to be sure, simplistic, but they do work, as has been proven time and time again. As a designer becomes more familiar with both his subject and his abilities he may want to diversify his results by adding retreats and variable results. CRT's such as those found in *Wellington's Victory* and *Tobruk* can only be tackled after extensive research as to firepower, casualty rates, etc. There are specific sources for such information (e.g., *Numbers and Losses in the Civil War)* but an overall familiarity with the subject is also required. New designers should refrain from tackling such subjects unless they have outside help. There is no substitute for experience in this area.

A good general precept to keep in mind is that the more tactical the game, the more definitive the CRT required. Large-scale strategic games such as *WWII* or *La Grande Armee* do not require the definitive complexity of, say, a *Terrible Swift Sword* or a *Torgau.* The CRT examples on the following pages may be of some help.

Readers will note in this sampler of tables that there are two basic kinds of CRT's (aside from the variations of complexity): the combat-ratio CRT and the combat differential CRT. In the combat ratio CRT, one side compares its total strength to the other and obtains a ratio in terms of odds (2-1, 3-1, 1-2, etc.). Thus one side is three times stronger than the other, or half as strong, etc. In a Combat Differential CRT the strength of one side is subtracted from the other; thus X is considered 4 *points* stronger than Y, rather than 4 times as strong.

There has been considerable argument over the relative merits of either of the two systems. Most prefer the former, it being the one they are used to. However, in certain situations (mostly tactical) the differential system has a great advantage. For example, in *Terrible Swift Sword,* Melee Combat, the relative strengths of the units were so similar that the chances of either side obtaining a ratio other than 1-1 or 1-2 was almost nil. However, differences in point *total* of 1, 2, or 3 points were common—and often telling. Thus, a differential CRT was chosen rather than a ratio.

TSS also uses a third-style CRT for firepower: a Point Total CRT. In such a CRT the strength of the defending units is not considered, in general. A player simply adds the total firepower being directed at a given target and, after adjusting for the type of cover the target has, he simply uses the appropriate firepower column to determine the results. This is another system that works best at a tactical level where such information can be gathered and used. At such a level, total firepower is more important than relative, whereas in a strategic situation, the de-

Figure 1: PanzerGruppe Guderian CRT's (final at top and prototype)

[9.5] COMBAT RESULTS TABLE

Die Roll	1-3	1-2	1-1	2-1	3-1	4-1	5-1	6-1	7-1	8-1	9-1	10-1	Die Roll
	Combat Odds (Attacker's Strength to Defender's Strength)												
1	A1	D1/A1	D1	D2	D2	D2	D2	De/A1	De	De	De	De	1
2	A1	Eng	D1/A1	D1	D2	D2	D2	D2	De/A1	De	De	De	2
3	A1	A1	D1/A1	D1/A1	D1	D2/A1	D2	D2	D2	De/A1	De	De	3
4	A2	A1	Eng	D1/A1	D1/A1	D1	D2/A1	D2	D2	D2	De/A1	De	4
5	Ae	A2	A1	Eng	D1/A1	D1/A1	D1	D2/A1	D2	D2	D2	De	5
6	Ae	Ae	A2	A1	Eng	Eng	D1/A1	D1	D2/A1	D2	D2	D2	6

COMBAT RESULTS TABLE (Panzergruppe Guderian Prototype)

Die	Odds											
	1-3	1-2	1-1	2-1	3-1	4-1	5-1	6-1	7-1	8-1	9-1	10-1
1	A1	D1	D1	D2	D2	D2	D2	DE	DE	DE	DE	DE
2	A1	A1	D1	D1	D2	D2	D2	D2	DE	DE	DE	DE
3	A1	A1	A1	D1	D1	D2	D2	D2	D2	DE	DE	DE
4	A2	A2	A1	A1	D1	D1	D2	D2	D2	D2	DE	DE
5	AE	A2	A1	A1	A1	A1	D1	D2	D2	D2	D2	DE
6	AE	AE	Ex	Ex	Ex	A1	A1	D1	D2	D2	D2	D2

COMBAT RESOLUTION TABLE

Die Result	1 to 4	1 to 3	1 to 2	1 to 1	2 to 1	3 to 1	4 to 1	5 to 1	6 to 1	Die Result
	Probability Ratios: Attack to Defense Strengths									
-1	AE	AE	AE	AE	AE	Ar	Dr	Dr	EX	-1
0	AE	AE	AE	AE	Ar	Ar	Dr	EX	EX	0
1	AE	AE	AE	Ar	Ar	Dr	Dr	EX	DE	1
2	AE	AE	Ar	Ar	Dr	Dr	Dr	EX	DE	2
3	AE	AE	Ar	Ar	Dr	Dr	EX	DE	DE	3
4	AE	AE	Ar	Dr	Dr	Dr	EX	DE	DE	4
5	AE	AE	Ar	Dr	Dr	EX	DE	DE	DE	5
6	AE	Ar	Dr	Dr	EX	DE	DE	DE	DE	6
7	AE	Ar	Dr	Dr	EX	DE	DE	DE	DE	7
8	Ar	Ar	Dr	EX	DE	DE	DE	DE	DE	8

Attacks at greater than "6 to 1" are treated as "6 to 1"

Attacks at less than "1 to 4" are treated as "1 to 4"

Figure 2: Simple CRT (*Napoleon at Waterloo,* SPI)

Figure 3: Variation of Simple CRT (*Battles of Bull Run,* SPI)

[8.0] COMBAT RESULTS TABLE *THE BATTLES OF BULL RUN*

Die Roll	*Combat Odds (Attacker's Strength vs. Defender's Strength)*										Die Roll
	1-5	*1-4*	*1-3*	*1-2*	*1-1*	*2-1*	*3-1*	*4-1*	*5-1*	*6-1*	
1	2/0	Ar	Ar	Dr	Dr	Dr	0/2	0/2	0/2	0/2	1
2	2/0	2/0	Ar	Ar	Dr	Dr	Dr	0/2	0/2	0/2	2
3	2/0	2/0	2/0	Ar	Dr	Dr	Dr	Dr	0/2	0/2	3
4	2/0	2/0	2/0	Ar	Ar	Dr	Dr	Dr	0/2	0/2	4
5	2/0	2/0	2/0	Ar	Ar	1/1	Dr	1/1	1/1	0/2	5
6	2/0	2/0	2/0	2/0	Ar	Ar	1/1	1/1	1/1	0/2	6

Odds less than 1-5 are treated as 1-5. Odds greater than 6-1 are treated as 6-1.

Figure 4: Moderate Complexity CRT (*Conquistador, SPI)*

[11.3] LAND COMBAT RESULTS TABLE

Die Roll	**Odds (Attacker to Defender Strength Points)**						
	1-3	**1-2**	**1-1**	**2-1**	**3-1**	**4-1**	**5-1**
1	A½/D¼	A¼/D½	D½*	D¾*	D¾*	De	De
2	A¾/D½	A½/D¼	A¼/D½	D½	D½	D¾*	De
3	A½	A¼	A¼/D½	A¼/D½	D¼	D½	D¾*
4	A½	A½	A¼/D¼	A¼/D½	A¼/D½	D½	D¾
5	A¾*	A½*	A½/D¼	A½/D½	A¼/D½	A¼/D½	D½
6	Ae	A¾	A¾/D½	A½/D¼	A½/D¼	A¼/D½	D¼

EXPLANATION: A=Attacker (takes loss shown); **D**=Defender (takes loss shown); e=All Units Eliminated; ¼,½,¾=Of Soldier Detachments Present, Portion Eliminated (rounding **upward**; see Case 11.21); *=Conquistador (1) eliminated if present. Odds less than "1-3" or greater than "5-1" are treated as "1-3" or "5-1," respectively.

Figure 5: Variation on No. 4 (*Eagles!* GDW)

DIE ROLL	**50%**	**75%**	**100%**	**150%**	**200%**	**300%**	**400%**	**500%**
0	T1/N	H1/N	H1/N	Q2/N	Q/N	N/Q	N/H	N/H3
1	H1/N	H1/N	Q1/N	Q3/N	Q/Q	N/H	N/H3	N/H2
2	H2/N	Q3/N	Q/N	Q/Q	Q/H	N/H3	N/H1	N/T1
3	Q3/N	Q/N	Q/Q	H/H	Q/H3	N/H3	N/H1	N/T1
4	Q/N	Q/Q	H/H	Q/Q3	Q/Q2	N/H1	N/T1	N/E
5	H/Q	H/Q	Q/Q3	Q/H2	N/H1	N/T1	N/E	N/E
6	H/Q	Q/Q3	Q/H3	Q/H2	N/T1	N/E	N/E	N/E
7	H/H3	Q/H3	N/Q1	N/T1	N/E	N/E	N/E	N/E

Figure 6: Specialized/Moderate CRT (*Desert War,* SPI)

[7.8] *DESERT WAR* COMBAT RESULTS TABLE

Attacks at less than −3 are not allowed; greater than +8 is treated as +8.

Die Roll	*Attack Superiority* -3	-2	-1	0	1	2	3	4	5	6	7	8	Roll
1	D1	D1	D1	D2	D2	D3	D3	D3	D3	D3	D3	D4	1
2	P	D1	D1	D1	D2	D2	D3	D3	D3	D3	D3	D4	2
3	P	P	D1	D1	D1	D2	D2	D3	D3	D3	D3	D4	3
4	P	P	P	P	D1	D1	D2	D2	D3	D3	D3	D4	4
5	P	P	P	P	P	D1	D2	D2	D2	D3	D3	D4	5
6	P	P	P	P	P	P	D1	D2	D2	D2	D3	D4	6

[7.81] EXPLANATION OF COMBAT RESULTS

P = Pinned; Affected units may not carry out that Game-Turn's plotted operation unless the operation was Fire Combat.

D1 = Disruption level one. Unit remains disrupted until Die Roll of 1, 2, or 3.

D2 = Disruption level two. Unit remains disrupted until Die Roll of 1 or 2.

D3 = Disruption level three. Unit remains disrupted until Die Roll of 1.

D4 = Disruption level four. Unit destroyed.

Disruption Effects: No operation plot may be executed or any new plot written while Disrupted. Disruption level effects are cumulative. Example: a unit with a current Disruption level of 1 suffers a Combat Result of "D2"; it now has a Disruption level of 3. "P" results have no effect on Disrupted units.

[7.82] HOW TO DISRUPT UNITS

Undisruption occurs in the Disruption Removal Phase of the Sequence of Play. All units that have a P marker on them have this marker removed and the unit is normal for the next Game-Turn. All the units with Disruption markers on them must roll a die to have the effect removed. Any unit that gets the necessary die roll has the marker removed and it is normal for the next turn. All units that do not get the right die roll remain Disrupted for the following Game-Turn.

Figure 7: Complex CRT for Specialized Strategic (*Frederick the Great,* SPI)

[9.6] COMBAT RESULTS TABLE

PERCENTAGE RATIO (Attacker divided by Defender)

Adjusted Die Roll	**33% thru 65%** Atk Loss %	Dfd Loss %	**66% thru 99%** Atk Loss %	Dfd Loss %	**100% thru 149%** Atk Loss %	Dfd Loss %	**150% thru 199%** Atk Loss %	Dfd Loss %	**200% thru 299%** Atk Loss %	Dfd Loss %	**300% thru 399%** Atk Loss %	Dfd Loss %	**400% or more** Atk Loss %	Dfd Loss %	**Adjusted Die Roll**
-2	25	5	20	5	25	10	15	10	10	10	10	25	•	P	**-2**
-1	25L	5	20	5L	25L	10	15	10L	10L	10	10	25L	•	P	**-1**
0	25	5L	25L	10	25	10L	10L	10	10	20L	10L	25	•	P	**0**
1	25L	5	25	10L	20L	10	10	10L	10L	20	10	25L	•	P	**1**
2	25	5L	25L	10	15	10L	15L	20	10	20L	10L	25	•	P	**2**
3	25	5	25	10	20	20	10	20	10	25	10	40	•	P	**3**
4	20	5	20	10	15	20	10	20	10	25	10	40	•	P	**4**
5	20L	5	20	10L	15L	20	10	20L	10L	25	10	40L	•	P	**5**
6	20	10L	20L	20	10	20L	10L	25	10	40L	•	P	•	P	**6**
7	20L	10	25L	25L	10L	20	10	25L	10L	40	•	P	•	P	**7**
8	20	20L	10L	10	10	25L	10L	25	10	40L	•	P	•	P	**8**
9	5	10	5	10	10	25	5	20	•	P	•	P	•	P	**9**

KEY: Number = % lost; L = Lowest ranked leader eliminated; P = Entire force captured; • = No Effect.

Figure 8: Complex CRT for Specialized Tactical (*Terrible Swift Sword,* SPI)

[10.9] FIRE COMBAT RESULTS TABLE

Die Roll	Total Fire Strength Directed into Hex: 1-3	4-6	7-9	10-15	16-21	22-27	28-33	34-39	40-45	46-49	50+	Die Roll
1	•	•	•	•	Pin	1	1	1	1	2	2	1
2	•	•	•	Pin	1	1	1	1	2	2	2	2
3	•	•	Pin	1	1	1	1	2	2	2	2	3
4	•	1	1	1	1	1	2	2	2	2	3	4
5	Pin	1	1	1	1	2	2	2	3	3	4	5
6(D)	1*	1*	1*	2	2	2	3	3	4	4	5	(D)6

Results shown are losses in Combat Strength Points; the Combat Strength is reduced by the number result.
(D)= If a "six" is rolled, roll again and check for Ammunition Depletion (see Case 12.21). **Pin** = unit is Pinned; see 10.84.
*=If Artillery is firing against Infantry-type, defending unit is **Pinned** instead of taking loss. • = No Effect

ADJUSTMENTS TO TOTAL FIRE STRENGTH COLUMN:
Never more than a **two**-Column final Shift in any one direction.

Move One Column to Left if...
...Defending unit is fired on through adjacent UpSlope Ridge hexside.
...Defending unit is fired on through adjacent Breastwork or Stone Wall.
...Defending unit is B-, C-, S- or P-type.
...Defending unit is in Woods hex.

Move One Column to Right if...
...Defending unit is limbered Artillery.
...Defending unit is Infantry in Column formation.
...Defending unit is fired on through an Enfilade hexside.

Move Two Columns to Right if...
...Defending unit is mounted Cavalry.

[13.9] MELEE COMBAT RESULTS TABLE

Die Roll		Melee Strength Differential (Attacker minus Defender): —4	—3	—2	—1	0	+1	+2	+3	+4	+5	+6	+7	Die Roll
1	Attkr:	R2	Eng	Eng	K	•	•	K	•	•	•	•	•	1
	Defdr:	•	Eng	Eng	C1R2	C1R2	C1R2	C1R3	C2R3	C2R3	C2R3	C2R3	C2R3	
2	Attkr:	C1R2	R2	Eng	Eng	K	•	•	K	•	•	•	•	2
	Defdr:	•	•	Eng	Eng	C1R2	R2	C1R2	C1R2	C2R2	C2R2	C2R2	C2R2	
3	Attkr:	C1R2	C1R2	R2	Eng	Eng	K	•	•	K	•	•	•	3
	Defdr:	•	•	•	Eng	Eng	C1R2	C1R2	C1R2	C1R2	C2R2	C2R2	C2R2	
4	Attkr:	C1R2	C1R2	C1R2	R2	Eng	Eng	K	•	•	K	•	•	4
	Defdr:	•	•	•	•	Eng	Eng	C1R2	C1R2	C1R2	C1R2	C2R2	C2R2	
5	Attkr:	C2R2	C2R2	C1R2	C1R2	R2	Eng	Eng	K	•	•	K	•	5
	Defdr:	•	•	•	•	•	Eng	Eng	C1R2	C1R2	C1R2	C1R2	C2R2	
6	Attkr:	C3R3	C3R2	C2R2	C1R2	C1R2	C1R2	Eng	Eng	K	•	•	K	6
	Defdr:	•	•	•	•	•	K	Eng	Eng	C1R2	C1R2	C1R2	C1R2	

EXPLANATION OF RESULTS:
Attkr: Effect on Attacker.
Defdr: Effect on Defender.
K=1 Strength Point becomes a casualty.
C1, 2 or **3**=That number of Strength Points is captured by the Enemy.
R1, 2 or **3**=Unit must retreat that number of hexes, then check for possible Rout.
Eng=The units are Engaged; see Case 13.83.
•=No result against that side.
Terrain and Leader Effects on Melee Combat: See Cases 13.3 and 13.4.
Retreats are always executed after losses; a C2R2 means remove two Points are captured **then** retreat the remaining units two hexes.

signer is more interested in the relative merits of the two sides.

With *PGG* (See Figure No. 1) the simple approach was chosen. The original CRT emphasized losses, and used a step-reduction system wherein German units, being better organized and better trained, were able to absorb losses more often. Thus a one-step loss would completely eliminate a Russian unit, whereas a German unit would simply be replaced by a unit of lesser strength. The designer has thus expressed the difference in cohesion and training between the two armies without going specifically into those subjects. This practice is called of integrating complex factors into simple mechanics *abstraction.* Certain information which the designer wants to get into the game as cleanly as possible is built into another section of the game for the sake of playability. Thus the *PGG* CRT not only covers relative strengths, it also covers such nuances as organization and training. These are tricks that most designers pick up with experience. Most games contain such abstractions.

One of the most famous abstractions in game design is in *Sinai.* There the designer (Jim Dunnigan again) built the air portion of the war directly into the combat strengths of the ground units (with the compliance of some special rules for Egyptian units) rather than go into the nuances of air-to-air combat as designer John Hill attempted to do in *Bar-Lev.* A great number of gamers failed to perceive this abstraction and wondered where the air war was. It was right in front of them; but, for the sake of playability, it had been factored into the game rather than presented directly.

In determining what to put into a game and what to abstract (and what to leave out!) the designer must be aware of his limits as well as the gamers'. The more he puts into the game, the more convoluted the play. And there is a point of diminishing returns here, a point that is often reached much sooner than most designers realize. Again, the choice is the designer's; it is his thesis and it is his intentions. If what he wants to show is exactly what happens in tank-to-tank combat—the various types of hits, armour effects, range capabilities, etc.—then he designs *Tobruk.* If he simply wants to ignore all the in-between steps and say the tank is either hit or it isn't, he does *Tank.* The approach of the two games is completely different; the end result (tank destruction) is the same. The difference—the designer's intent.

After making basic decisions concerning movement, combat ability, initial deployment, etc., certain choices must be made about the system itself. What would make the game unique, what would make it accurate, and, above all, what would make it interesting? In three key areas, Dunnigan came up with design decisions that directly affected these questions: Overrun (a form of combat), Supply, and an idea that had been introduced in a previous game—*Invasion America*—Untried Units.

At exactly what point in the working of the design such decisions are made is sometimes unclear, but they usually occur as initial testing proceeds. In the case of the Untried Units, the decision was made to use such a system from the beginning. Dunnigan, in his research, had realized that the capabilities of the Russian units varied greatly; furthermore, even the Russian generals did not know how their units would react when they reached the front lines. Some divisions literally melted away at the first sound of the guns! The best way to simulate this would be to take an average strength for all units and then vary that strength—using the average as a median—so that there is a large number of units with differing capabilities. Then, the strength would be placed on one side of the unit/counter, while only the basic unit type would be on the reverse. The player would thus not know what strength the unit had until it actually entered combat!

The Untried Unit system *made Panzergruppe Guderian.* Yet, it was *not* an original idea—although it was unoriginal only in that it derived from another game. Designers should never be afraid to borrow concepts from other games where those concepts apply to the system they are using. It is in such a way that the state of the art advances. An idea from one game is used in a second; it is refined and honed and reapplied under different circumstances until it becomes a standard rule. The novice designer should resist the temptation to be innovative at every step of the way; this is flatly impossible. There is very little done in wargaming design today that is purely innovative. Most innovation is simply sagacious adaptation, the ability to fill a need with a previously used rule. In its previous incarnation in *Invasion America,* the Untried Unit rule was simply an added fillip for spice, used to represent unknown strength of militia-type units. In *PGG,* the rule defined a style and a system; it was refined from its initial application and graduated from novelty to standardization.

Such a design innovation does not come about immediately; certain things have to happen. First of all, the need must arise. With Untried Units, it was the need to represent the idea of the "unknown", the unknown representing lack of information for either side on what a unit would do in battle. But that need would not have led to anything had not a simple change been made in counter production: printing the counters on both sides! This may seem a fairly simple thing, but it was not until quite recently that such a production step would be readily undertaken. (The cost for such a step is approximately 1¢ per counter sheet at the production level.) The reverse side not only gave designers more "counter space", but it provided them with a field for imagination, a field that was filled as the need arose.

Thus, the above example is used to show what kind of atmosphere produces "new ideas". They do not simply arise because the designer wants a new idea; they arise when there is a need to fill and when the capability to fill it is there. For *PGG,* Dunnigan quickly realized that Untried Units would be perfect for what he had in mind, and he set out to use them as the base on which to build his game.

Dunnigan, as designer, does a lot of counter pushing when he designs. That is, he has his first-draft map, his initial set of counters with their values, and a hazy idea of what he wants to accomplish. He then puts the units on the gamemap and begins to play with them to determine what needs to be done. He jots down the main ideas that occur to him on the crucial aspects of the game, and he then repeats the process to see if they fit. To be sure, other designers do not operate in this manner. Some like to think out many of the game rules before attempting any play, others do something in between. The novice designer will most likely have a good idea of what he wants to do, and it is best to get to the initial testing stage as soon as possible—to see what is at least working.

At this point Dunnigan wrote out, in short paragraph form, his basic ideas on the structure of the game. He had his map, his counter-mix (which had to be 200 or less to fit the format), his CRT and his basics. His job was finished; the rest was up to the developer.

It is at this point that the novice designer must change his approach to a game. He has the information and he has the basics. His development will now take two divergent courses: he will, in some instances, start to refine and remove; in other cases he will be adding rules. The difficult part is deciding when to do what. And that is where playtesting comes in.

With *PGG* "in-house" playtesting began as soon as the initial developer, Kip Allen, began work on the game. The developer plays the game with another member of the staff to get a feel for the system and some general reaction. This goes on for several days, or until the developer feels he can proceed. Usually notes are taken and some ideas jotted down. It is important to keep a record of what works and what doesn't, as well as any ideas that occur, for it is nearly impossible to remember everything you want to do without some form of notes. Items that are slated for removal, but are overlooked simply because there was no reminder of such, tend to remain in the game, reinforcing their position until the developer no longer recognizes them as part of the problem.

Initial testing of *Panzergruppe Guderian* revealed several problems, one of which was that the game did not work as history: i.e., the results the testers were getting did not coincide with the parameters of historical happenstance. The Germans were alternately going nowhere at the beginning and then running hogwild over the entire gamemap later on. Furthermore, the Soviets, if they did not get wiped out immediately, were counter-attacking—and winning—quite early on! The Soviet player did not seem to have any idea of what he should do; his fate seemed to be entirely in the hands of the German player.

These were flaws that should have been spotted immediately. Unfortunately, Allen was in the process of relocating and could afford little concentration on this, his last game. The obvious solutions were overlooked, no research was done, and Allen tried to work out the problem by simply testing the game over and over. It was time for a change.

With time growing short, the game was handed to Richard Berg. Berg had designed only two games previously (one of which was yet to be finished) and had little experience with WWII subjects. But that supposed weakness proved to be a strength as Berg looked at the problems from their roots.

His first step was to question the gamemap, which seemed to be channeling the game along non-historical lines. Dunnigan admitted that he distrusted his original map sources, so a new map was done. This time, with the help of SPI Librarian Ginny Mulholland, several different maps of the area were found. It took two weeks to combine the maps into one test version, but the final product was very different from the original. Novice designers, unfortunately, rarely have the facilities to make such changes, e.g., the new *PGG* map entailed extensive camera blow-up work, as well as overlaying one source map with another. Such processes are usually out of the reach of all but the professional designer.

Amateurs must content themselves with the painful process of doing everything by "eye and hand". But it is important to remember the ability of your game-map to direct the flow of the play of the game. Perhaps if the game is not going as planned and you have examined all other aspects, you should check your map and its sources again. Terrain can be adjusted (even "fudged", if you want to be so bold) to meet your needs, but this must be done through sources. Do not underestimate the power of the map.

The new *PGG* map now had quite a different outlook. Its forest areas—based on aerial photographs of the Soviet Union—was now dispersed, rather than lumped in the four corners of the map. This, along with a more realistic river network, produced the watershed/plateau effect that was so crucial in the Smolensk area. The German thrusts were now centered—and directed—into the areas historically chosen. The terrain now assumed the same importance and effect as it had in the original campaign. With the general trend of the battle now proceeding along historical lines, Berg turned to the outline of rules presented to him by Dunnigan.

The first major change was wrought in the CRT. In the original CRT, the Russian player suffered severely in combat. Since combat was mandatory when units were locked into Zones of Control, a one step loss meant total elimination of one unit for the Russian player when attacked by the German. Furthermore, since ZOC's were rigid, the Russian player was forced to counterattack—usually at poor odds—on the following turn, resulting in even greater losses. The result: smashing German victories and total Soviet frustration, from a player's viewpoint. After some initial and fruitless attempts to change the ZOC rules, Berg perceived the problem as one in which the Soviet player was forced to conduct a static defense. He had to stay and fight; he could not choose to delay and retreat—a favorite Soviet tactic. The problem was the CRT, and with only one or two seemingly minor adjustments, a new CRT was developed.

The Soviet player was now given the choice of *either* losing a step (or unit) or retreating all units in the stack. This gave the Soviet player (as well as the German player) a strategic *defensive* choice, something that had not yet been used regularly in wargaming. (A somewhat similar, if cruder, system was used in *Korea* back in 1969). *Exchanges* were replaced with *Engaged's,* a somewhat more subtle form of the former, and split results were added to give greater divergence in the outcomes. This rather simple adjustment (it took less than a half-hour to work out) produced a complete change in play. The Soviet player could now *choose* his mode of defense; play was back in his hands.

Such a development decision was, as you can see, a process of trial and error, test and retest. The designer was of course consulted, and he agreed with the change (a basic one) as it reinforced the operational tactics of the era. And this is how most development work is accomplished; constant working and reworking to see what happens. Try to visualize the game as a modular set of rules, rules that you can take out and replace with another set to see whether it is that particular section that is at fault. In such a way the developer can see what portions of his game are working and what sections are not. That makes it all the easier to replace and rewrite non-working sections.

Various other sections of the rules were now tackled, as problems arose, and an insight into the mechanics that produced the changes might be enlightening as to how the process of development works.

UNIT STRENGTHS

As playtesting progressed, it became apparent that the Soviet units had too great a capacity for counterattack. German units were loath to close with their Soviet counterparts, a circumstance which had little basis in history. However, the Soviet strengths worked well for Soviet defense. Thus only one portion was out of balance while everything else seemed to work. The answer was to use an old method: split the combat strengths on the Soviet units. Thus, while the German units retained a one-digit combat strength, the Soviet units now had an attack strength and a defensive strength. The attack strengths of Soviet units were reduced (overall) by approximately 25%, with the Untried unit system allowing for a great deal of individual variation. The resultant variety in Soviet units proved especially pleasing to the players as well as productive of a more historical result.

GERMAN DIVISIONAL INTEGRATION

The original rule, as written by Dunnigan, stated that ". . . German regiments are doubled in *attack* strength if stacked with *another* regiment from the same division." Although this produced the desired results, it did not satisfy the historian in either the developer or the designer. Testers constantly asked why the stacking rule wouldn't effect defense equally. Moreover, the rule had been placed in the game to reflect the increased strength that German regiments had when operating as a whole division, with support and back-up capabilities. The rule was thus developed to include defense in the integration rule. It then further changed to reflect actuality by requiring that *all* units from a given division would have to stack together to attain the benefit. Once these changes had been implemented, German play underwent a corresponding change: German attacks became more local and more cohesive (as they were historically) and specific breakthroughs began to appear. The game took on an increased "blitzkreig" atmosphere because of the increased necessity for concentration. Thus, here is an example of an historian's hunch backed by testers' questions producing a slight variation that added significantly to the historical flow of the game.

OVERRUN

More trouble occured with the overrun rule than with any single rule. In the first place, overrun is often a difficult concept for gamers to understand. It is combat, yet it is movement.

Furthermore, the rule as originally written produced some strange results and seemed arbitrary at best.

> "*OVERRUN*—Units may overrun defending units during either one of the Movement Phases. The cost will vary—if +2MP are expended, the attacker attacks at ⅓ Strength; if +4MP are expended, the attacker attacks at ½ Strength. Combat actually takes place in the Defender's hex. If attack is successful, the attacker may move on provided it is not in any other ZOC. If attack is unsuccessful, the attacker must stop in the last vacant hex along his movement path."

The original overrun rule was an attempt to use German momentum to create breakthrough situations. Thus a player was given an opportunity to attack during his movement phase to push units out of the way. The player would sacrifice some of his mobility, and, because he was attacking on the move, some of his strength. Problems then arose with the question of moving into and out of ZOC's. Players could never understand what they could do.

The first development was a definition of movement into and out of ZOC's *vis a vis* Overrun. This proved to be quite difficult, and it took several rewrites before a relatively clear rule was formulated.

> [6.56] When conducting an Overrun attack, Players may ignore the Zones of Control of units which are exerting influence on the hex from which the Overrun attack is coming. That is, they may occupy the target hex if the Overrun is successful regardless of other Enemy ZOC's. However, if a Friendly unit finishes an Overrun in an Enemy Zone of Control, it may not **move** any further in that Movement Phase; however, it may conduct another Overrun on an adjacent hex providing it has the necessary Movement Points.

Once this was cleared up it became a question of balancing the strength of the German overruns with the desired result. After several permutations and combinations of effects on the original rule, it was found that halving the total strength at a cost of 3 MP's worked best. This is a form of development that is best described as "legwork". The developer must maintain constant supervision of the playtesting to see what is working and what is not. The effects of such a subtle change as outlined above cannot be discerned by spot checks only; the developer must be able to see what is happening at all times.

Of course the amateur designer/developer has no problem with this. He is usually present (possibly the *only* one present) at all stages. He is so close to the game that he usually can spot such changes and make adjustments. Here again, familiarity with both topic and game help immensely. Don't be afraid to make adjustments, but make sure you follow up on such adjustments to see if they have the intended effect.

ADDITIONS DURING DEVELOPMENT

Several rules were actually added during development. Of course the Victory Conditions were not defined until the end of playtesting, when it could be determined what was happening in the game. Throughout playtesting, the players always had a rough idea of what they were supposed to accomplish. The original victory conditions were quite simplistic: "Germans: get 3 mechanized divisions off the east edge of the map in supply. Russians: hold the Rzhev-Vyazma-3830 highway under control." Not only did these instructions reflect the discarded original map, they also did not quite reflect the historical situation. Gradually it was determined that the German player would have to take and hold the major cities (in supply), and this became the basic condition. The German player received points for the various cities on the game-map, graduated as to importance. This total was affected by the ability of the Soviet player to eliminate entire German divisions as well as to recapture lost cities. Further adjustments were made according to the number of reinforcements the Soviet player diverted from other (off-map) areas. The result was a simple point-count system which corresponded to the historical situation.

Actually, the Victory Conditions as written reflect a certain bias in the game. The Soviet player would be hard-pressed to keep the Germans from obtaining a marginal (and historical) victory. This bias reflects the game as a simulation, and is something that new designers should keep in mind. To balance a game absolutely 50–50 is impossible (even chess isn't so balanced). In *Panzergruppe Guderian,* the play balance favors the German. However, for the German player to really "win" a game, he should have to do better than history. The designer can thus "have his cake and eat it", as it were. The Victory Conditions and levels of victory in *PGG* reflect both a game win and an historical win. Equal emphasis is given to both, and the game thus functions well on both levels. This cannot be done in every instance, but it is something to keep in mind because gamers do not all play for the same reasons. Some play to win, some play to recreate history (with myriad positions in between). Victory Conditions such as those in *PGG* satisfy both of these groups. It is an enviable position to attain.

Victory conditions themselves are not *that* important in terms of game design. In most instances they will formulate themselves. You don't want overly complex conditions requiring endless calculations. But the conditions should be obvious to the players from the start of the game, so that they know what their objectives are. Conditions should be so concise and clear as to avoid any possible confusion—and concurrent inertia—on the part of the players.

Another rule that was introduced late in the development stage was Interdiction. This was done more for flavor than anything else. It enabled players to simulate, on an ultrasimplistic level, some of the air war of the Eastern Front. Interdiction also affected supply and enabled the players to make strategic decisions on this area as well. Thus another decision was given to the player while keeping the game fairly simple. This is a trick to watch out for. The best games are the ones that keep the rules relatively simple but give the players a fair number of decisions to make. This gives the players a greater sense of participation and relieves the theory of "prisoner of the die". Such an effect was achieved with the new CRT (as was explained above), and designers should be on the lookout for similar items.

The above exposition concerning the history of *Panzergruppe Guderian* is useful not in that it shows how *PGG* came about, but rather in that it throws light on the design/development process. There is no way that someone can tell you how to be creative. The only way to learn is through experience. By using the lessons of *PGG,* the novice designer can perhaps gain some insight into the mechanics of design and development.

In subsequent sections, the various areas of design and development will be discussed on a more random and discursive level with an eye towards giving the designer ideas, rather than specific instruction. Designers should feel free to do what they want; yet they must understand that if they want to produce a professional-looking and appealing product they will have to use a professional approach.

SECTION III:

Image and System

Graphics and Physical Systems Design

by Redmond A. Simonsen

INTRODUCTION

One of the least discussed and least understood aspects of conflict simulation design is, ironically, that which is most obvious: the graphics and physical systems that make a game a reality in the hands and eyes of the gamer. In fact, the better the graphic design, the more likely it will *not* be noticed. Since, in game design, the overriding mission of the graphic designer is to *communicate* the substance of the game to the user, heavy-handed or flashy images that call attention to themselves (rather than their message) are actually detrimental. The type in which this book is set is a simple example of this: each letter is well designed and crafted—and yet, when strung into organized arrangements (i.e., words) the individual letters become invisible. If the typeface was eccentric or exotic in design it would be hard to read and would detract from the message rather than convey it.

I intend to deal with the *terra incognita* of graphics with regard to the following areas:

1. The role of graphic design in simulation games.
2. How the professional works and how a game is produced.
3. How the amateur game designer can produce a more professional looking product without the resources available to the professional.

Aside from the practical matters of production and technique, much of what I'll be presenting is naturally my own personal "doctrine" developed over the past twelve years (seven of which I've spent working on conflict simulation design). Since there is no "one-true-way" in matters of art, my graphic philosophy would doubtless be challenged by other artists and/or gamers. I would like to point out (at the risk of sounding elitist) that that which is superficially eye-catching may in the final analysis, not be the best solution for a game/graphics problem—and many non-artists have difficulty in separating that which looks good from that which works well. The two are not mutually exclusive—but neither are they necessarily mutually *inclusive*. Readers familiar with my work are aware that I am an advocate of form-following-function and will accept my statement that the graphic "philosophy" expounded here is not a mere rationalization of an artist's quirk of style, but rather a sincere explanation of a point of view developed and put into practice over a number of years.

THE ROLE OF GRAPHICS IN THE DESIGN OF SIMULATION GAMES

More than almost any other type of game, simulations are enormous information processing and learning problems. Even the simplest game requires the player to manipulate dozens of discrete pieces (units) in hundreds of possible cell locations (typically hexagonal); sort out thousands of relevant and irrelevant relationships; and arrive at a coherent plan of action (a move) several times in the course of the play of that game. It is a testament to the power of the human mind that anyone can begin to play such complex systems let alone do it well. The average gamer may have several dozen game titles in his library, each of which differs from the other—yet miraculously he can sit down on any given night and (with perhaps a glance or two at the rules) play a creditable game.

Given this large burden on the player, the challenge to the graphic designer is clear: make the information the player uses clear, organized, accessable, and pleasing to look at for long periods of time. To use a military metaphor, the player is an unspecialized demolitions man defusing a complex bomb and receiving instructions on how to do so via a radio. The game is the bomb, the game designer is on the other end of the radio and the artwork *is* the radio. If the radio is faulty, the unclear signal may break the concentration of the demolitions man (with unpleasant results). Now the qualities of a good radio are fairly obvious: good signal-to-noise ratio; adequate range; reliability; and good design of human factors (ease of handling, etc.). Metaphorically, these qualities translate fairly well into the qualities of good graphic design in games—what is not so clear, however, is exactly what constitutes a good signal-to-noise ratio in graphics or just what value to place on "reliability" (which translates as consistency of format). And although the gamer is not vaporized when faulty graphics causes him to "detonate" the game he's playing, the fact that it has indeed turned out to be a "bomb" is certainly unpleasant. Virtually every gamer has had the experience of struggling through what might be an otherwise good game, hampered by the fact that the organization and design of the components prevents him from easily understanding what he is about—and thereby losing concentration and interest in the game.

Many of the factors which degrade the performance of the graphic system in a game are not obvious to the average gamer. Some of these considerations are technical in nature: e.g., the size and style of type used on the counters; the intensity, surface quality, and range of colors used on the maps as they affect vision; the weights of the lines used to separate sections of charts or forms, etc. Others are organizational: the tables and charts should be well-integrated and logically formatted; the terrain symbology should be a development of a consistent approach; the rules should be presented in a systematic, accessable format, etc. Although they may not be consciously perceived, these factors all add up and impact upon the user as he plays the game. Wrong design choices can conspire in such a subtle manner that the gamer may not be able to pinpoint *why* the game is troublesome but he'll be aware that *something* is wrong and is preventing him from getting the most out of the game.

It is sometimes difficult to separate poor (or good) *graphic* design factors from poor (or good) *game* design factors. There is a great deal of feedback between the two. Of course, no matter how good the graphics and physical system, they cannot turn a weak game design into a strong one (although they can sometimes cosmetically hide an inadequate game design, at least for a while). But the reverse is possible: bad graphics and poor physical systems *can* ruin a good game.

Before going further, I should perhaps explain what I mean by the term "physical system". The term is really my personal jargon for the graphic engineering of game elements. The more graphic engineering the artist can build into the game equipment and rules, the easier and more enjoyable becomes the play of the game. Examples of this are: the Production Spiral used in SPI's *War in Europe* game system; Turn-Record Tracks with built in information on special events; Phase Records that are themselves diagrams of a complex sequence of play (such as in SPI's *Fast Carriers*); game maps with the set-up printed directly on them; integrated combat results tables (with terrain effects built in). A good physical system is characterized by its organization of game information to such an extent that the presentation actually accomplishes some of the "work" of using the raw information. It is possible (and often is the case) that a game is well-designed graphically, but no serious attempt at physical system design is evident.

Decoration Versus Design

One of the major questions a graphic designer must answer when approaching a new game project is: how much decoration is necessary, desirable, and allowable within the context of the information problem the game presents? The term "decoration" is used to indicate those graphic elements which have no practical bearing on the utility of the components. Well, if decorative elements are non-essential, what's the point of putting them in at all? In most games, some decoration is psychologically necessary in order to create the proper mood for the player who is going to spend several hours engaged in a game. Properly used, decoration helps the player to relate his activity in the game to the historical activity being simulated.

Unfortunately, many artists concentrate most of their efforts on these decorative elements and virtually ignore the practical requirements of the game. Decoration is information—*unnecessary* information—which if present in overabundance distracts the player from the truly important, game-play information he must have. Most people are, unfortunately, easily impressed by highly decorated games. Only when one plays such a game several times does the realization come that all that cute stuff is actually hampering play. Of course, the best possible combination is a well-designed physical system which has an overlay of just the right amount of mood-enhancing decoration.

Usually, the more complex the game-system, the *less* decorated it should be. When counters carry several different values and symbols; when the terrain is highly varied, when the mechanics of play are very involved, it is then that decorative effects should be kept to the bare minimum. It could almost be stated as a quasi-mathematical theorem: decoration varies inversely with complexity. As you might expect, it could also be said that the amount of graphic engineering required varies *directly* with the complexity of the game-system. Any such statements must always be qualified with the comment that there are no hard and fast rules in art. There will always be exceptional cases and mitigating circumstances. Even the borderline between practical and decorative graphic elements can sometimes become fuzzy. For example, tank silhouettes on tactical game counters may not be strictly necessary, but they do help the players to quickly distinguish AFV's from other types of units. Regardless of how simple the game might be, however, there are some elements of decoration that I am dogmatically opposed to. First on my list of such elements is the placement on maps of extensive terrain that has no effect on play whatsoever. There's nothing sillier than (for example) a large swath of desert glaring at the player when that desert is no different from ordinary clear terrain. If the designer wishes to impart the fact that there's a desert on the map, he could much more reasonably place a simple line of type indicating the name of the desert.

Second on the list are orders of battle that go strictly by historical designation without giving the player the option to ignore the designation and set up the game and the reinforcements purely by unit type and value. Particularly in large games, the set-up is one of the most tedious and time consuming exercises that a player must endure, every effort should be made to make this process as painless as possible.

Another such mistake occurs in counter designs which use large flag symbols (for example) to display nationality (when a simple color change is all that's necessary) and the important numerical data is squeezed into the small remaining space. In this case, as in many others, it's really a matter of proper emphasis being ignored or subordinated to some eccentric concept of "historical flavor". There's nothing *wrong* with such flavoring—it's simply a matter of knowing how much salt to put in the soup.

THE GRAPHIC DESIGN AND PRODUCTION OF SIMULATION GAMES

What follows is, in game jargon, the sequence of play of designing and printing simulation games.

I. THE PRELIMINARY GRAPHICS PHASE

A. Game Designer does sketch map using standards and procedures established by Graphic Designer.

B. Game Designer writes rules outline and develops order of battle.

C. Game enters developing and testing. Game Developer consults with Graphic Designer concerning particular problems in game that can be solved via graphics.

II. GRAPHIC DESIGN AND FORMATTING PHASE

Finished sketch map, counter manifest, and rules are turned into Art Department Copy Editor.

1. Rules are searched for bugs and inconsistencies.
2. Graphic designer designs terrain symbology to be used on map.
3. Graphic Designer designs counters, choosing type-styles symbology, and layout of various types of units.

III. GRAPHICS EXECUTION PHASE

A. Typesetting Machine Operator sets counter values and map labels.

B. Copy Editor and Graphic Designer format tables and charts; TMO sets material.

C. Boardman pastes-up counters.

D. Boardman does artwork for map/color separation.

E. Boardman lays-out and pastes up rules.

IV. FIRST CHECK PHASE

A. Game Developer proofs counter sheet and map.

B. Copy Editor and Game Developer proof rules.

C. Corrections performed by TMO and Boardman.

V. INSPECTION AND MARK-UP PHASE

A. Graphic Designer inspects finished, corrected art and mechanicals.

B. Graphic Designer marks-up work, giving printing and color specifications.

VI. FIRST EXTERIOR PRODUCTION PHASE

A. Artwork sent to camera-house (to be photographed and rendered as large film negatives).

B. Negatives sent to printer to be stripped up as a series of film negatives in preparation for plate-making.

C. Printer makes paper checking copy of film negatives and sends these to Graphic Designer.

VII. SECOND CHECK PHASE

A. Graphic Designer, Copy Editor, and Game Developer proof checking copies.

B. TMO and Boardman perform necessary corrections and correction film patch is sent to printer.

VIII. SECOND EXTERIOR PRODUCTION PHASE

A. Printer corrects negatives, and makes plates.

B. Components are printed and sample press sheets sent to Graphic Designer.

C. Graphic Designer and Printer give Binder and Die-Cutter, instructions on final finishing.

D. Countersheets are mounted and cut by Die-cutter. Rules, maps, and charts are folded, bound and cut by Binder.

E. Finished parts are partially or wholly collated by Binder and returned to Publisher. Game ready for sale.

Of course, the foregoing sequence is that which is followed at SPI—other publishers will have their own variations of the same basic steps. SPI is vertically organized, i.e., all major production steps except the actual printing and binding are done by SPI staff. Other publishers may have one or more of the phases in the sequence performed by outside services—for example, typesetting or box art. In fact, some publishers have the entire graphic/printing sequence done for them by other organizations. While this approach has some advantages in staff salary savings, it does limit the amount of control that the publisher has over the finished product. When a game is produced totally "in-house" as at SPI, there is more opportunity for the creative people involved to interact and produce a better product. The elements of simulation games are extremely complex, and, therefore, the greater the communication between game designer, developer, graphic designer, and production staff, the greater the chance of producing an integrated, coherent whole. The four main parts of the typical game each present their own characteristic problems in design and execution. The graphic designer must be aware of these basic problems in the context of the individual game system for which the components are being designed. Although they must all work together, the parts of a game can be examined as distinct problems: These parts are: 1) the map; 2) the counters; 3) the rules; 4) the charts, tracks and tables that operate in the game.

The Map

Usually, the game map represents a specific section of the world's geography in terms of those of its features that have a bearing upon the maneuver and combat of military units. In other cases, the map is a synthesis of typical terrain designed to provide varied situations for a group of scenarios (the latter type of map is most often found in tactical level games).

The first step in map design is up to the game designer. He determines the scale and the features that will be shown on the map. The game designer or developer then prepares a sketch prototype of the game map. If the game uses a hexagonal grid, he does this by placing the source map on a light box and then placing the blank hex-grid paper over it. He then sketches in the terrain, employing a set of standards previously developed by the graphic designer. If there are unusual or particularly significant features on the map, a consultation with the graphic designer will be necessary. This sketch map (and copies of it) will be used throughout the testing and development stage of the game.

A final sketch of the map is prepared for submission to the Art Department and the graphic designer and the game developer discuss the requirements of the game as they have grown and changed in the course of development. The graphic designer will "interrogate" the game developer and designer as to what features of the game can be built into the map in order to facilitate play. This process is referred to as "interrogation" only half in jest: many times a developer or designer is so close to the design that he cannot imagine the need for various graphic aids to be incorporated into the game-map system. The graphic designer (who should of course be basically familiar with the game) can often draw out of the developer/designer important pieces of information that can be successfully integrated into the map design. What follows is a partial list of such questions:

1. Can the basic set-up be printed on the map using unit-pictures or codes?
2. Can the victory conditions be expressed on the map by coding the cities or sites that may be the objectives?
3. Would it be useful to code entry and exit hexes or reinforcement sites?
4. Are there any seasonal/weather changes that can be displayed on the map without interfering with the basic terrain?
5. Are there any rules other than victory conditions, that make some terrain feature or site important enough to warrant a graphic emphasis?
6. If the game involves the production of units,

are there any values or devices that can be built into the map to aid the player?

7. If the sketch map indicates more than one terrain feature in a hex, which takes precedence (and can the map be rationalized so that there is only one feature per hex)?
8. Are there any superfluous terrain features on the map or are there any redundant features that can be eliminated to clarify the actual, operative terrain analysis?
9. What are the effects of the various features? Is there a natural hierarchy that can be expressed graphically?
10. Are there any games in print which use a similar or identical terrain system? How well does that prior system serve the present need?

Other questions will suggest themselves in specific design situations—there is no magic formula for creating a map that is not only pleasant to look at but which, more importantly, serves and supports the game system.

There are, of course practical considerations that must be taken into account when producing a game map in a commercial environment. Limitations of the printing process, time restraints on creating the art, the price of the game and the cost of the components, the size of the packaging, etc., are some of the factors that influence the design of the map (or indeed, any piece of artwork). An effective graphic designer, however, can turn these limitations into advantages (or at worst, an operable discipline) and produce a map that works better and is better looking than one produced without restraints by a designer who fails to treat the map as a total unit to be integrated into the game system.

At SPI I've set up a number of standards for game maps in their various stages of production. This has been done for practical, management reasons but it also serves to reduce the learning problem for the player in the final, printed version of the map. In the sketch stage, the designer/developer makes use of standard colors and symbology developed by me in order to regularize the production of map prototypes. This enables to designer/developer to quickly produce his map without having to "re-invent the world" each time a new game is done. The standard symbols available are comprehensive enough to cover most of the variations of scale and period found in conventional conflict simulations. Because of the use of these standards, playtesters and production people alike can more readily read the map and copies of it can be re-drawn with less difficulty. Many standard approaches are also used in the final production versions of the maps. As mentioned, this relieves the player of some of the learning problem (if he's familiar with the games in the same general family) and also saves time for the printing people and for me. Because we do so many maps at SPI, these standard approaches have ample opportunity to experience refinement and evolution. The maps we do today are strikingly different from those that we did three or four years ago (although some basic resemblances can be detected if one looks hard enough).

When starting a map, the designer/developer may have a specific source-map in hand or may have to search for one. Usually maps are drawn from several sources. The Art Department often plays an advisory role in this initial stage. It's always best to use as a primary source, a map that is close in scale to that which will actually be used in the game. If the source map is more than three times smaller or larger than the game map, problems of accuracy and readability will arise in the sketch stage. The scale of the source map can easily be related to kilometers-per-hex by using the scale ratio found on almost all maps. For example, a map of Europe might be in the scale 1: 6,000,000. This means that every one unit-of-distance on the map equals six million of those units in reality. Since the size of our hexagons is known (in millimeters) we can just convert this known size into kilometers and multiply by the right-hand number in the ratio. The standard hex size in most SPI games (and non-SPI games) is 16 millimeters, i.e., 0.000016 kilometers. Thus:

$0.000016 \times$ map scale = kilometers per hex.

The Europe map mentioned previously would then yield a game map with exactly 96 kilometers per hex if used same-size. A different scale is achieved by taking the source map and photostatically reducing or enlarging it to the desired scale. For example, if one wanted a map with only 32 kilometers per hex, the source map would have to be enlarged by a factor of *three.* To arrive at the necessary map reduction or enlargement, one need only divide the scale of the source map by the desired final scale. For example, if we wanted the Europe map to be 53 kilometers per hex, we'd divide 96 by 53 yielding 1.81, meaning that the source map would have to be photostatically enlarged to 1.81 times its original size. These simple formulas can be worked in reverse if one *knows* the reduction or enlargment necessary (to fit a map on a standard hex sheet) and wishes to discover the scale of the resulting map.

Once the map is scaled the designer or developer traces the black and white photostatic copy of the map using a light-box (simply a even source of light under a flat frosted glass that allows one to see through ordinarily opaque paper). This is one of the most tedious and time consuming initial steps. It has to be carefully done or else errors will be introduced at an early stage that may remain in the game through its publication. One of the most typical mistakes in this stage is to position the map incorrectly on the hex field. Because humans are creatures of convention, many people automatically orient the source map to the hex field in a rigid "north-is-straight up" fashion. Many times even the professional overlooks the possibility of rotating the map orientation to yield a better fit that results in more useable map per hex sheet and fewer dead-spots in the playing area. A good example of economical, unconventional map orientation is the *War in Europe* map system. If the same map were imposed in a strictly orthogonal fashion, it would have consumed twelve or more map sheets rather than the nine it actually takes. When dead spots occur unavoidably, the developer must inform the graphic designer of the opportunity to use the space on the map for some other map-related function. It's also pointless to execute terrain that is never going to be used—not to mention misleading to the player.

Once the map is aligned within the hex field and prototypes are sketched, the game begins to undergo playtesting. Items on the map will sometimes be modified or eliminated and new items may be added. It is important to maintain a control on the sketch maps so that it is always known which is the most recent and most "official" version. The final map sketch, submitted to the artist for translation into finished form, must be as accurately drawn and as complete as possible. As mentioned, the developer and artist must confer at the turnover point in order to treat any special problems and/or develop other game-features which can possibly be built into the map.

The typical game map is a simplied physical map overlaid with a political map. The graphic designer must make the proper choice of colors and symbology to create a map which will have high utility for the player and yet be pleasing to the eye. In my work at SPI, I've developed a number of approaches to the many terrain representation problems that confront me regularly. Some of the more basic and routine problems I consider to be "permanently" solved—and I apply that "permanent" solution to that type of terrain in almost every instance. For example, I almost always handle rough terrain as a continuous tone of a certain tint of brown totally filling the hexagon. In other instances, I have a number of standard approaches that I draw from based upon a judgement of appropriateness to the game scale and subject. This standardization allows me to concentrate on the many totally new terrain/map problems that I regularly face (as a result of the tremendous number of different game systems produced at SPI).

The graphic designer has available to him a range of choices as to how to convey a given type of terrain or map element. These divide into categories which I'll now list in order of their recognition value (i.e., the ease with which the average person senses the presence and meaning of the graphic element).

1. Color and tone
2. Shape and pattern
3. Symbol
4. Typography and outline
5. Position

What this means is that those elements most essential to the interpretation of the map should be represented by change of field color—since humans with normal eyesight most easily recognize differences in color. The problem with this is that there are a limited number of colors that will be instantly recognized without having to closely compare them to the other colors in the group used. For instance, if one chose to use four different types of blue (all meaning different things) it would be difficult for the average person to discriminate precisely amongst them unless they were all placed closely together for comparison. Colors also have the characteristic of *apparently* changing when the neighboring colors change. Place a green next to a red and it looks greener. Put an orange-yellow next to a red and it will look yellower. Tonal values also apparently change relative to the adjacent tones—light tones look lighter next to dark tones and vice-versa. This means that as a practical matter, the graphic designer can effectively employ only about three changes of value within a color and must limit himself to the use of no more than four or five colors (all of which should be spectrally well separated). Overlaid on these limitations are the limitations of the printing process. SPI maps usually use only *three* standard ink colors (although the visual impression is that of many more colors).

Additionally, the more colorful a map is the harder it is to read in an overall sense: the patchwork quilt of a multi-colored map can be confusing to the eye and tiresome to look at for long periods of time. For the same reasons, use of raw primary colors should be avoided in map work except as accents. When using color to convey information, the designer must strike a balance between the ability of the gamer to separate with his eye the difference in color and the harmony of the color scheme. In order to eliminate some of these difficulties, I've chosen to print almost all SPI maps on a paper-color called Sandstone—this color automatically harmonizes the ink colors printed on it and also reduces the glare problem. Incidentally, it's a basic principle of mine that no map should ever have a white field. The most common mistake in the use of color on wargame maps is too make the colors too harsh and bright and to surround them with large expanses of white paper. Not only is the effect produced ugly and hard to look at but it also is suggestive of a childlike level of presentation that undermines the legitimacy and seriousness of the game map.

As logic would indicate, the best type of terrain to represent as a field of distinct color or tone is that type of terrain which is itself areal in nature and can be associated mentally with one of the colors available to the artist. Obvious examples are fields of blue for bodies of water; brown for rough or mountainous terrain; green for woods; yellow for desert, etc. In conventional geography, color tints are used to separate one nation or political subdivision from another. This option is usually not open to the game map designer, since to do so would mean that the colors of the terrain would be altered by the field color. There are exceptions to this of course, and one of my own game designs, *After the Holocaust* is such a case. The subdivisions were very important and the terrain limited and relatively insignificant, therefore color was used to separate the many subdivisions on the map.

After color, the eye best perceives differences in pattern or texture. There are a number of topographic patterns that lend themselves very well to game map applications. Some of these are: foliage patterns suitable for wooded areas; swamp patterns; stipple patterns suitable for rough and mountainous terrain; cross-hatching useful for man-made areal effects. These patterns are produced using prepared adhesive sheets (manufactured by a number of graphic arts suppliers). Depending upon the number of ink colors available to the graphic designer the use of these patterns expands the effective number of combinations of color and tone by a factor of two or three. Care must be exercised however, since the application of too many different patterns can give a map such a "busy" texture, the user's eye will be disturbed and distracted.

Patterns and textures assume even greater importance when the designer is limited to the use of only one or two ink colors. In one-color work, the patterns and textures, in combination with photomechanical tints, actually pinch-hit for the unavailable colors. Theoretically, it's possible to do *any* map in one color (i.e., black or dark brown ink) using only patterns for areal effects.

Such a solution of terrain can be a useful exercise for the designer even if he has more color available to him, since if he can solve the problem in *one* color he can then gain valuable insights into how to enhance his solution with the use of additional colors. This is analogous to executing a detailed charcoal sketch of a subject before going to paint it in oils.

Shapes are allied to patterns and texture as carriers of information. The organic, puffy edge of a patch of forest clues the eye very quickly. The splashy form of slopes and ridges and the irregularity of land masses are other examples of how the shape of large terrain features help to identify them for the gamer. Symbols can be thought of as smaller, more organized shapes. In game map design, symbols are most often used to characterize a "point" feature—something that resides in a single hex or location. Such things as cities, resource centers, industrial sites, forts, railheads, airfields, and ports are examples of terrain feaures that can successfully be represented by the use of symbols. Symbols are usually pictographic, i.e., they actually look like stylized versions of the feature they represent—or they are simple drawings of objects associated with the feature being represented—for example, a resource center might be represented with a pick and shovel symbol. Non-pictographic symbols are used when the feature being represented has no obvious object with which it is well associated or when the number of other symbols on the map calls for the use of abstract symbols to avoid confusion. Stars, for example, might be used to denote capital cities or arrows to indicate invasion hexes.

When using symbols, the designer must remain conscious of the fact that too many symbols, or symbols that lack recognition value, may actually confuse the player rather than convey the information. Moreover, symbols suffer from their trait of being obscured by the counters occupying the hexes containing the site being symbolized. This, incidentally, is an important consideration regardless of terrain treatment—how much will the counters affect the visibility of the playing surface? One solution (which I often use) is to fill the hex with the feature so that even when its occupied, the terrain is still visible around the edges of the playing pieces. This gives the map a somewhat more abstract appearence—but I feel that the sacrifice of naturalism is worth the additional utility gained by this technique.

By changing the color and/or size of the symbols, more variations can be achieved if truly necessary. Symbols can be combined with each other to form ideographs that convey more complex messages than any one symbol could. For example, a map shows three types of installations (ports, fortifications, and airfields) each of which must be characterized as being "major" or "minor" and also be identifiable as to which player possesses them originally. One could use twelve different symbols, but a better solution is to use a symbol in a circle to indicate a "major" installation and a different color to show ownership. This way by using only one more symbol (in conjunction with three basic installation symbols) and one color change one creates a simple system that is easy for the player to remember and easy for the eye to spot on the map.

When designing symbols to be used on maps, some basic considerations must be kept in mind:

1. The number of different symbols should be kept to a functional minimum. Don't make arbitrary distinctions between items that, in the game, are treated identically. For example, if all fuel resource sites are operatively the same, don't show petroleum sites as little oil wells and coal sites as little picks and shovels. Instead, use a common symbol that evokes the "fuel" concept rather than the irrelevant fuel type.
2. To be effective, symbols must be simple and well designed. A complex, cluttered symbol does not contribute to player information retrieval. Most symbols are best treated in silhouette form.
3. The symbol should be evocative of the basic concept of the thing for which it stands. The test of a good symbol is how well it is understood *without* recourse to a key or legend. Whenever the artist is doubtful of the recognition value of his symbology he should show them to an associate without telling him what they mean, and ask that person to quickly interpret the symbology.
4. The symbol should reproduce well in the map environment. Even if the symbol is effective in isolation, unless it works in the context of the map, it can be a bad symbol. Also, when several symbols are used, they must all work well together. They sould have a consistency of style and approach that makes them into a total system.

The most abstract and generalized set of symbols that the designer routinely employs is the alphabet. Typography is an extremely useful tool in map design when used with restraint and a good appreciation of the reading problems presented by game maps. Beyond the usual typography of maps (place names and hexagon numbers) the designer can use type to carry more information than the simple words themselves. Changing the size of the type, for instance, can be used to express relative importance of cities. Change of style can be used to indicate differing categories of terrain. Changing color can associate the type with a given player or with a particular type of terrain. The designer must be careful not to use too many different styles of type on the same job—doing so creates visual confusion making the map harder to read. A good rule of thumb is to use no more than two families of type, each with their roman, italic, bold, and bold italic variants. The choice of exactly which typefaces to use can be made from a broad spectrum of possibilities—there are hundreds of different styles commonly available.

The two basic typefaces that I use in my work at SPI are called *Times Roman* and *Univers*. These were chosen because, first of all, they are well-designed typefaces and also because they are very versatile and legible whether used in text sizes or as headlines. This book in fact, is set in Times Roman. Although it is classified as a modern typeface, Times Roman looks appropriate when used on any subject from classical to far future. Univers is the name for a very large family of typefaces with many different weights and widths available. I avoid the more extreme members of the family except for very special applications. *Strategy & Tactics*, SPI's flagship magazine is set in Univers (specifically Univers 45/46 and 55/56—there are so many Univers variants they are *numbered* instead of named). More will be said about typefaces in the rules and counter discussions.

Positioning of map elements can also be used to send information (or to merely reinforce information). This must be done very carefully however, since its easy for the gamer to misinterpret signals that rely totally on their relative position. For example, if pictures of the playing pieces are printed on the map to indicate the starting set-up, one can discriminate between one force and another simply by rotating one set of images 90°. Or, for instance, if there are two sets of numbers associated with every city on the map (victory points and supply points, for example) one number group could be placed north of the cities and the other to the south. Such a system can cause a problem, however, if players are sitting on all sides of the map. As much as possible, type should be placed at right angles to both players so each has an equal chance of reading it. It can be a hidden advantage to one player if he sits on the side of the map where all the copy is right-reading from his viewpoint.

The graphic designer must avail himself of the foregoing techniques to create an efficient and handsome map. He must also treat the map as a total problem rather than trying to defeat-it-in-detail. He must, early on, make a basic decision as to how *naturalistic* his map treatment will be. That is to say, how much like an aerial photograph will his map look as contrasted to the opposite extreme of being very abstract and "game-like". A number of considerations bear upon this decision. For instance, the scale of the game should influence the degree of naturalism: the more tactical the map, the more it should seem like an aerial view of the actual terrain. There are exceptions to this rule—particularly if the game is very complicated or has unusual mechanics that can be simply expressed by the use of abstract graphic devices (such as hexside bar-coding). Then too, there are tactical games that have "all-purpose" terrain, i.e., terrain that changes with the scenario. In such cases, naturalistic terrain is actually a hindrance even though it's appropriate to the scale.

Strategic level maps, almost automatically, take on the appearance of traditional geographer's abstract. On extremely small scale maps (for example, SPI's *Global War)* the necessity to represent the terrain in terms of hexagons forces even greater abstraction in the artwork. [Terminology Note: the *smaller* the scale of the map the larger the area it covers]. In *Global War,* I designed a system which allows the terrain to be represented with considerable accuracy and without any doubt as to the nature of a given hex or hexside. The coding system used received mixed reviews, the negative part of which I feel comes more from the strange appearance of the map rather than the ease of use (which is high). The system uses three different types of hexside bar-codes to indicate whether or not a given hexside is impassable to land units, sea units or both types. In addition, all coastal hexes are coded with a hex-filling tint. The ghost of the actual shorelines is printed over all this for reference purposes (the literal coastline is ignored). This system has tremendous advantages over a conventional treatment of the map problem. It avoids any dispute as to the character of a given hex. It clearly shows which movement paths are open to a given type of unit. It even allows the geography to be presented accurately (without having to distort landmasses to conform to the hex pattern). The disadvantages are: difficulty and expense of execution in both test and final art stages; player resistance; and interference of coding with other necessary terrain features.

The use of hexside coding was something I've been experimenting with for as long as I've been designing. Back in 1970, I made one of my first uses of it in SPI's *Bastogne* game. Later, in *PanzerBlitz,* I sophisticated the system somewhat by using several different bar colors to code the hexsides. In many subsequent games, hexside codes became standard practice to represent blocked hexsides or any special characteristic of a hexside that could not easily be expressed in terms of the relationship of the areal terrain covering the hexside. Some day I'll give into a secret urge of mine, and do an accurate map of some battlefield rendered solely as hexside codes and full-hex symbology (not having any resemblence to naturalistic terrain at all). One could probably get the best playing map that way (but I'm afraid that the typical wargamer would rebel against such a coldly mathematical presentation).

SPI pioneered the unmounted game map—first out of economic necessity, then later out of conviction—it became the standard format for

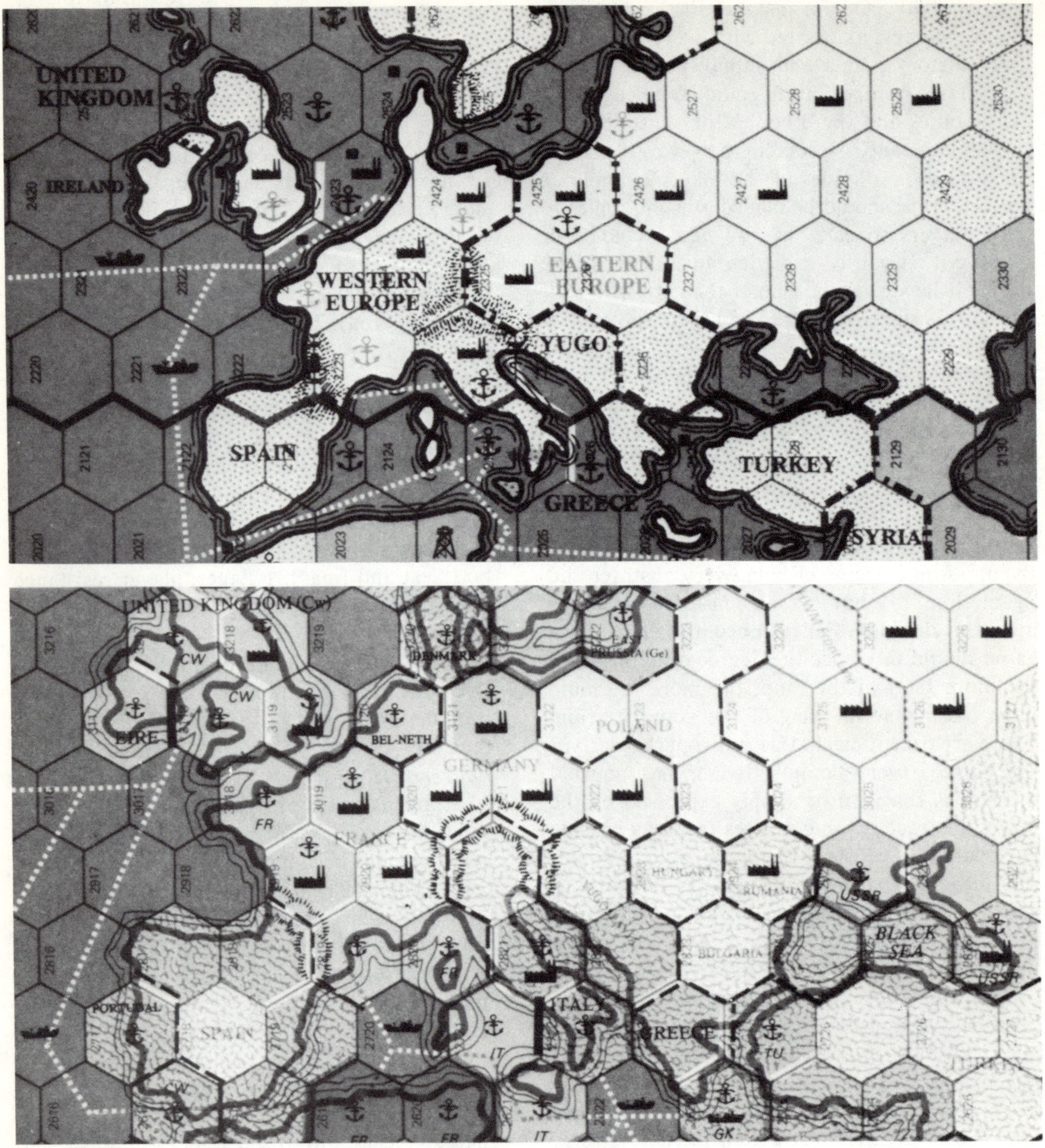

Two maps of the same geographic area done in similar scales for two different games (top: *World War 3*; bottom: *Global War*). The *Global War* map is the more successful of the two (and indeed was a development of the *WW3* map). *Global War* represents a high-level refinement of the blocked hexside system used to express complex terrain relationships in the most unequivocal manner. By contrast, the WW3 system, even though it looks clearer, suffers from ambiguous land/sea relationships and distorted geography.

playing surfaces in simulation games. Traditional parlor games (and the first commercial wargames) all employ the typical glossy-paper-glued-to-cardboard game board. This is fine for parlor games (*Scrabble* and *Monopoly* and the like), but it causes some problems for wargames. First of all, wargame maps are big—typically about twice as big as the boards for parlor games. This bigness requires that a number of splits and seams be created in the mounted board in order to fold it up into a manageable package. This causes an annoying interruption in the playing surface plus it creates registration problems from one panel to the next—this registration (i.e., matching) problem can become critical in a hex-grid wargame. In most production lines it's impossible to mate the map sections from the same printed sheet. This means that the west portion of the map is cut out of one sheet; the central portion or portions, from another; and the east portion from yet another. This increases the matching problem (and adds the bonus problem of maintaining consistent color from one sheet to the next). Additionally, mounted maps make the game expensive (see The Business of Wargaming later in this book). The cost of the really large games so popular today, would be astronomical if all the maps were mounted.

The usual defense for mounted maps is that they are more durable and more convenient to set-up and use. In surveys of its customers and informal tests, SPI has discovered that *un*mounted maps actually survive more playings, in better shape, than mounted maps! This is less surprising if you consider that the stress placed on the joints of mounted maps is very great whenever they are packed and unpacked. Due to the trend towards compact packaging, most mounted wargame maps come in two (or more) completely separate sections that must be butted together to form the whole map. These sections easily separate by accident during play. Unless they are stored in perfect conditions, even the best mounted maps gradually warp creating more discontinuity in the playing surface. If this represents convenience of set up and use—make mine vanilla. Traditional mounted maps make use of glossy paper. I can't think of a worse surface to stare at for three to six hours of game playing than glaring, glossy, varnished inks printed on coated white paper. The use of glossy paper (and the press varnish that is laid over it) is simply an appeal to the more childish weaknesses of the consumer—"bright and shiny is good". Unfortunately, even the mostly rational intelligent consumer group that makes up the wargaming audience is susceptible to this appeal to some degree. Much of it is based on conditioning and tradition. Many people *expect* a game board to literally be a "board" and to be shiny and bright—so when one panders to this expectation the people are superficially pleased even though they are *not well served.*

This is not to say that *unmounted* maps are free of problems. The slight belly some unmounted maps exhibit causes a "trampoline" effect when the center of the sheet is pressed. This can be avoided by proper back-folding of the map and by using a few pieces of masking tape to stretch the map tight to the table—a few minutes of trouble that hardly is significant measured against hours of playing time. Better still, a thin sheet of clear acrylic plastic can be laid over the map to provide continuous stable surface (alternatively GE's *Lexan* clear plastic is even better than acrylic—it's virtually indestructable and will not support combustion). Using plastic re-introduces a glare problem—but this can be solved by using matte fixative spray on the plastic. The use of a high-surface (but uncoated) earth-colored quality cardstock allows SPI to produce good-looking, glare free, one-piece game maps that suffer none of the disadvantages of mounted maps and only have minor drawbacks of their own. As the new generation of wargamers grows up in an SPI environment (and as older gamers acclimate to unmounted maps) the unmounted, non-glossy mapsheet grows in acceptance and desirability.

To be fair, it should be pointed out that SPI "caved in" and made concession to the tradition of mounted maps in its line of hardboxed games that it distributes mainly through toy and department stores. This was more a matter of getting the product past the wholesale buyers than appealing to the final customer. Chain-store buyers are even more tradition-bound than the customers they serve. The buyers care more about the box cover and the *weight* of the game as a selling point than anything else. The *heavier* the better—and mounted maps are heavier.

The "perfect" game map surface would combine the characteristics of both mounted and unmounted maps: it would be rigid; one continuous piece without splits; fold to compact size yet open perfectly flat; have a homogenous cross-section; and be truly durable. As yet there is no such perfect surface that can be made cheaply enough to be commercially viable. There is some promise though in the new plastic laminates that are coming into the stream as replacements for paper in certain applications. Until some designer (I hope it is I) comes up with a better solution, the gamer will have to cope with the less than perfect surface for this all-important component.

The designer should never lose sight of the fact that most gamers are deeply influenced by the game map: a good map goes a long way towards creating a positive impression of the game. Since the map is the most constantly used component, it should be the most effective in doing its job of providing the basic environment for the game.

The Counters

Typically, the wargame playing piece, or unit counter, is a half-inch square of cardboard that represents a military unit. This little chip of paper and ink can be the most concentrated source of information in a game. The numbers and symbols found on the counter (in conjunction with the probability tables) are the lifeblood of a manual simulation.

A wargame counter starts its career as a set of values arrived at by the game-designer (and/or developer) through research, empirical testing, and some inspired guesswork and approximation. The game developer makes a trial set of playing pieces by drawing them on blank counter sheets using a felt-tipped pen. During testing and development, the order of battle and the values of the counters will undergo modification (sometimes to the point of eliminating whole categories of values and/or units). Before the counter manifest is turned in to the graphic designer for production, the developer usually consults the graphic designer as to how the values should be formatted in the manuscript and what, if any, are the peculiar problems associated with that particular counter mix. The developer and the artist must discuss the quantities and types of counters necessary in order to insure that sufficient pieces are available to meet the needs of the scenarios and that the total limit is not exceeded.

The typical counter has a standard military symbol dominating its upper half and two large number values filling the space of its lower half. It may or may not have a set of smaller numbers associated with the symbol (i.e., those numbers that are its historical designation). This "standard" counter has been used so often that it's almost unnecessary to do any real graphic design work on such a counter mix. The great number of new games developed by SPI, however, have introduced the need for many new approaches to the problems of counter design.

To design a successful set of counters the graphic artist must be familiar with the game system and its requirements. As in map design, there are always opportunities to place some less-than-obvious piece of information on the counter in order to simplify play. In counter design, however, there is a premium on space and an imperative for legibility that severely narrows the options of the designer.

Most wargame counters are the size they are because of the limits of map size and packaging. Hexagons are the most common grid-cell—given the size of the hexagon there is a maximum practical counter size. For any size hexagon (measured from side to parallel side) one can discover the theoretical maximum counter size by multiplying the diameter of the hex by 0.866. Applying this formula to the standard 16mm hex, one comes up with the maximum counter size of 13.86mm—which is about 1mm larger than the actual size of the standard counter. The counters *should* be a millimeter or two smaller than maximum to allow them to comfortably occupy adjacent hexes.

The actual working area available to the designer is actually less, of course, than the full counter size. There is an unuseable border area on each counter called the *safety*. Items which are printed in the safety area are in danger of being cut off by the knives of the steel die used in the counter manufacturing process. The die-cutting firm used by SPI requires a safety of 1.6mm. (slightly more than 1/16th of an inch). So on a 12.7 by 12.7 counter face the designer must confine his images to an area measuring 9.5mm square (about 3/8ths of an inch square). This area is called the *live area* of the counter face. If the counters are back-printed (i.e., double-sided), the limitations are even greater on the backside—the live area of the back is reduced by another millimeter all around.

It might be useful to explain here how counters are manufactured. SPI does the artwork for its counters at 150 percent of the final size (to make it easier on the artist). This art is photomechanically reduced to proper size, rendered as a film negative. The printer takes the film negative and makes a plate from it. The image is printed on good white paper about the weight of the pages of this book. This printed sheet is then sent to the die-cutter who mounts the sheet on cardboard and then places it into a die-cutting press to strike in the cuts that form the individual counters. As the reader can see, there are a number of steps involved all of which require accurate *registration* (i.e., image alignment) of the counter sheet. Each registration step introduces a slight error factor—this is the reason for the safety requirements previously referred to. Even with a safety, a number of sheets will be improperly die-cut. In the not too distant future, there may be such a thing as *laser die-cutting* which will increase the accuracy of the process by a whole order of magnitude. 'Til then we must cope with the present limits of technology.

Given the limits of the process, the graphic designer must strive to produce the most useful counter image. Counters should be designed with an information hierarchy in mind. This is simply a categorization of items to be displayed on the counter according to their relative importance:

1. Who owns the counter?
2. What type of counter is it?
3. What is the primary value(s) of the counter?
4. What historical or functional information not included in the above categories is necessary for the play of the game?
5. What historical information not included in the categories above is desirable to display on the counter even though the information is not functionally necessary?

Usually, *ownership* is signaled by the field color of the counter. When one player controls counters or several different nationalities or subgroups, the color field may be changed to reflect those groupings—thereby making one graphic device carry two pieces of information. The *type* of counter is most often indicated by use of a symbol—either the standard military symbol or a special symbol specifically designed for that counter set. *Primary values* (i.e., strength and movement ability) are most commonly shown as large numbers. Secondary *necessary historical information* is usually shown as either a smaller symbol or an alphanumeric set in type 75 to 50 percent of the size of the primary values. *Non-functional information* can print as either small type or very small symbols. Such information should never be allowed to dominate the counter—permitting it to do so is one of the most common mistakes of counter design.

Another basic question that the designer must answer is: what is the information load of the counter and is it appropriate to the game system? Traditionally, the designer attempts to put as much useful information as possible on the counter face. It may be possible, however, to eliminate some information as redundant. It may also be possible—and desirable in specific games—to pull the information from the counter and place it on a data sheet separate from the playing pieces.

The counters in many air warfare games are examples of unit-information removed from the counter and otherwise displayed. SPI's *Foxbat & Phantom,* for instance, uses the counter only to show ownership, position, and altitude—all of the remaining information is carried on a separate aircraft control sheet which itself uses a number of markers to display changes in the basic data—such as speed, climb, missiles remaining, etc. The feasibility of such an approach is not always obvious—the graphic designer must put himself in the player's place and make judgements concerning this basic question.

As a general rule, the more tactical the game, the more information will be displayed on the counter; the more strategic, the less information. If, however, a game becomes *very* tactical an information threshold is passed which demands that data be *removed* from the counter (as in the example of the air games). One might say that the extremes in scale result in very simple counters and the middlegrounds produce the most variation and problems.

After one decides how much information to place on a given style of counter, the artist must determine what means are to be used to display the various categories of information. What follows is a list of this means as they apply to the design of counters:

Color: Variation in the color field is almost always used to distinguish between nationalities. Sometimes within a nationality, special groups of units may warrant their own distinctive color field. For example, if airborne units are particularly important, they might be differently colored. Games that use leader counters often have these counters accented in colors different from the combat formations. In this way, color is being used to distinguish unit type. In rare instances, color is used to signal the value of a piece—in the game *Sorceror,* I used a color-hierarchy in the combat and movement systems. Whatever the use the color is put to, it should be remembered that a counter is a very small patch of color for the eye to distinguish—if the different colors used are too similar the eye will not be able to discriminate between them except when they are immediately adjacent to each other. The terrain colors should also be kept in mind—I regret to say that more than once I've been guilty of making counter colors so close to some terrain colors that the counters disappear when placed on the map—camouflage is fine in real war but a definite hindrance in simulated conflict. Dark colors also tend to look alike—particularly under the less than ideal lighting conditions of most game playing. Such colors also have the disadvantage of making the type on them hard to read unless the type is reversed out of the field (i.e., produced as a white image) which causes some difficulty in printing.

Symbols: For most military units, there exists a semi-abstract symbol developed by the US Army for use on maps and tables of organization. These symbols are perfectly adequate for most games (see the section in this book on symbology). Experienced wargamers know most of these symbols by heart, so when the designer uses them he reduces the learning problem for the player (although I should mention that new players are sometimes mystified and intimidated by this symbol system). In some games, however, these symbols are inappropriate and the designer must resort to either making up his own abstract symbols or using simple silhouettes of the basic element of the military force represented by the counters—for instance, a tank for an armored unit or a cannon for an artillery unit. Particularly in older period games, there is more reason to use silhouettes as opposed to standard military symbols—it just seems more appropriate. Then too, if the counter is *very* simple, a silhouette helps to dress it up and prevent it from being too sterile. Each method of indication has its adherents in the hobby. My feeling is that neither is clearly superior to the other and that the designer must rely on his judgement to guide him into the right choice for a particular game. Whatever style the symbol, the more complex it is the bigger it must be in order to remain readable on the small counter surface. Silhouette-type symbols are most often found in tactical level games, and in air and naval games—one reason being that the abstract symbology for such games is not as well-developed as that for land units.

Type: This is perhaps the most important means for communicating information on the counter. The majority of wargames present the values of the units as digital information and just how those digits are arranged and displayed on the counter is critical. The *size* of the type is the first consideration: the designer must make sure that when the counters are put into play, the players can see the numbers on them! Virtually as important (but often not well selected) is the typeface. As mentioned in the discussion on maps, I use Univers and Times Roman typefaces almost exclusively. Particularly for modern games, the face Univers 65 is especially well-suited. Its numbers are compact and clean and easy to read in small sizes. When space is at a premium, Univers 67 (a condensed style) can be substituted. An equally well-designed face is Helevetica Medium (although this face sets a little wider than Univers—which is why I don't often use it). Most gamers would find it difficult to tell one sans-serif typeface from another—and my choice of styles may seem somewhat moot. On a subconscious level, however, the readability of the right typeface wins out and the gamer will find that the counters with the right typeface are easier to read than those using the wrong selection—even though he would be hard-pressed to explain exactly why. Type sizes are measured in *points* (of which there are 28 per cm or 72 to the inch). This book is set in 10 point type. This does *not* mean that the letters are 10/72nds of an inch high—it means that the "natural" distance from baseline to baseline of this type measures 10 points (in this book two additional points of space have been inserted between the lines to improve readability—this is called setting 10 on 12). The actual capital letter height of 10 point type is about two-thirds the nominal point size, or 6.6 points. The normal reading distance of a wargame counter is 30cm to 100cm (about a foot to a yard). If the type on the counter is smaller than 9 point, most people will have difficulty reading it at the far end of the working distance. As an experiment, place this book on a table and stand about one meter away. Can you easily read the following numbers?

45-2 889 0-21 4 67-6 90 5-2 0-9

The numbers above are 9 point Univers 65. You should be able to read them at one meter—but if they were much smaller you'd have a lot of trouble doing so.

But it's not enough to use the right size type and typeface—the designer must also properly position and space the type so that the gamer has a fair chance of reading the counter. This can be quite a problem when the counters have several different values on them (as for example the counters in SPI's *Sixth Fleet).* When the counters require elaborate historical designations the graphic challenge is even heavier. Usually such material is set about half the size of the primary values—which means that unless the gamer is an eagle-eye, he must get fairly close to read this information. I have a somewhat radical viewpoint on historical designations on counters—I think they should be *eliminated* and replaced with an easy to read code—and I'll say more about it later in this section. In the foregoing discussion of type sizes, the reader should keep in mind that the actual art for SPI counters is done at 150 percent of the final size—so I actually set my counter type larger than the sizes mentioned to account for the reduction to final size.

A cross between type and symbol is the use of codes on counters. This can be as simple as a single bar of color to distinguish the "back" of the counter from the "front" or as elaborate as multiple dots and dashes to separate the counters into several discrete groups (for example, by stacking ability or morale values). The use of such coding relieves the designer from having to put another number on the counter and saves space by allowing him to superimpose one piece of information on another (since simple bar codes can be run right over the type without any ill effect).

As of this writing, not much use has been made of differences in *shape* as carriers of information. Most wargame counters are the same simple squares of cardboard for all types of units. The reason for this lies mainly in the expense of building new steel dies to have such counters manufactured. Then too, there could be some handling problems with differently shaped counters in play on the same map. But assuming these difficulties can be overcome, there are a number of interesting possibilities that have yet to be seriously explored by the major manufacturers. Take for example, a hypothetical strategic game with air, land, and sea units. The ground combat units could be traditional squares; the naval units rectangles; the air units circles, etc. Even the size of the counters could be varied somewhat to reflect relative importance (this is a very old concept which doesn't find much application in wargaming). There are a few games around that make some attempt at this—but no really thoroughgoing design work has been done. It might be useful here to shatter a myth that some gamers put their faith in—and that is the supposed "ideal" nature of hexagonal shaped counters on a hexagonal grid map. There are a number of flaws in the hexagonal counter concept: 1) A hexagon shape affords much less useable space for images on the counter. 2) Hexagonal counters are actually harder to handle and stack on the map. 3) Hexagonal counters so fill the hex they occupy that it's hard to tell *what* kind of terrain is in that hex; 4) A hexagonal die is very difficult to construct accurately—which is responsible for even greater irregularity than normal.

Regardless of the shape of the counter, there is the question of *surface*, i.e., matte finish as opposed to glossy finish. If you've read my earlier treatment on maps you know that I strongly prefer matte finish counters. The angle of incidence of the gamers line of sight with respect to counters in play varies from 90 to 20 degrees. If the counters have a glossy finish then at any given time during play, some of his counters will be unreadable due to glare. To my mind, the dubious benefit of slightly more brilliant color reproduction is not worth the true drawback of glare. I don't feel that I have to appeal to the wargamer's susceptibility to the "shiny syndrome" in order to produce a good looking set of counters. SPI ran a test to see how much difference glossy counters made to gamer acceptance: in issue number 58 of *Strategy & Tactics,* half the subscribers were sent glossy counter sheets and half were sent matte finish counter sheets. The feedback survey in the magazine asked them to rate the game—and a separate question asked them to rate the counters. On a zero to nine scale, both groups of readers rated the acceptability of the game *identically!* The ratings of the counters differed by only half a point (the glossy counters getting a marginally higher rating than the matte counters). In conversations I've had with gamers who actually *play* the games rather than just collect them, most have indicated that SPI's matte finish counters are better to play with (in fact as a result of the test mentioned above, I got a number of angry letters accusing me of "copping out" and caving in to the pressure for shiny counters). Those manufacturers who do produce glossy counters like to hoot about their durability compared to matte counters. This is simply not true—properly printed matte counters done on high quality stock with properly dried inks are actually *more* durable than glossy counters. This is because on an uncoated sheet, the ink actually sinks into the surface and bonds with the

EXAMPLES OF PLAYING PIECES (from various SPI games)

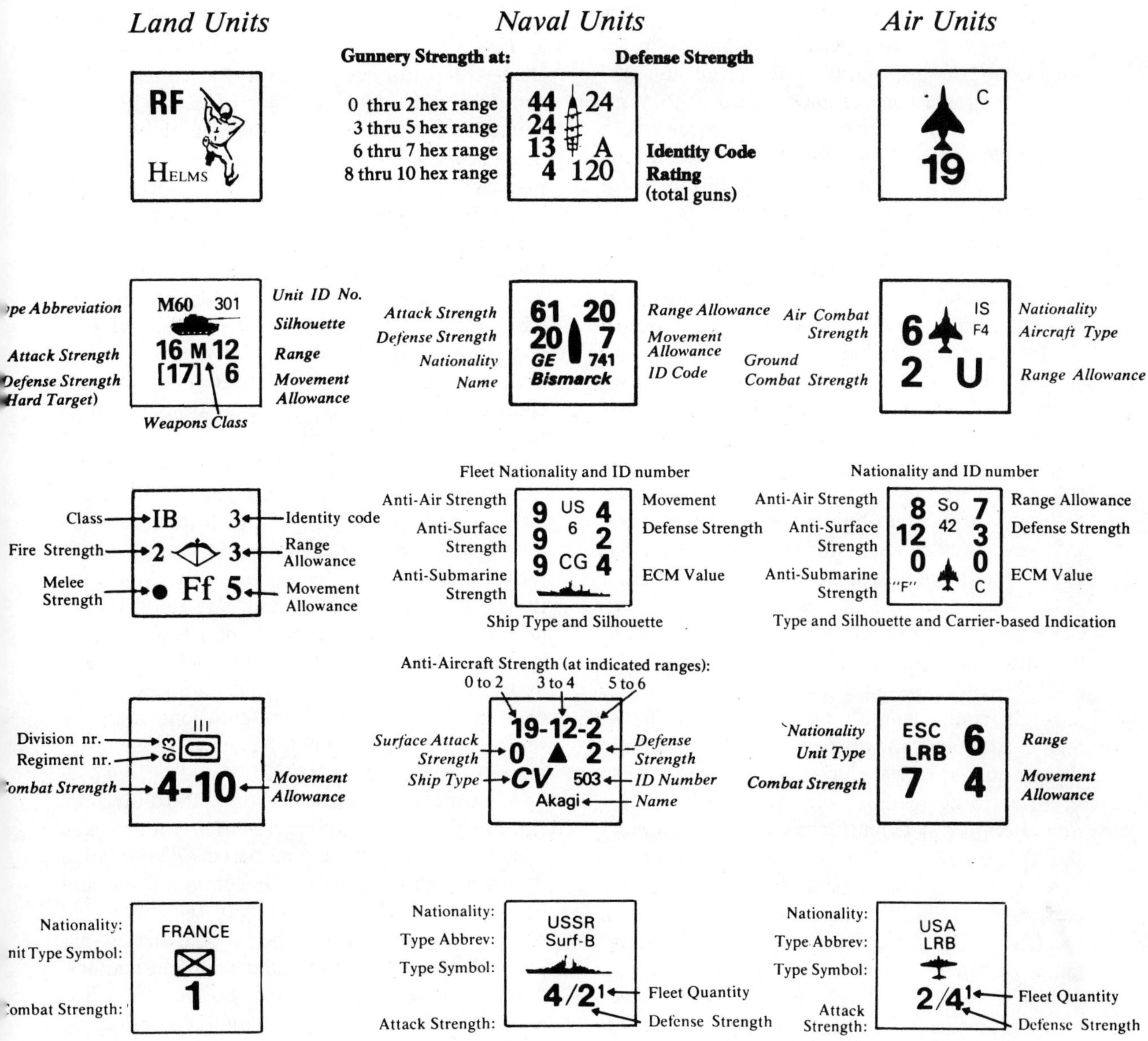

The above samples are arranged in three columns (by general type) and in ascending order of scale reading down the columns (i.e., from tactical level games through strategic). Land column: *Patrol, MechWar 77, Prestags, Panzergruppe Guderian, Global War.* Naval column: *Frigate, Dreadnought, Sixth Fleet, Fast Carriers, Global War.* Air Column: *Foxbat & Phantom, Oil War, Sixth Fleet, Invasion America, Global War.* Note that in general, the low level and high level counters are relatively simple, whereas the middle level tactical counters are more complex. Note that the naval units are something of an exception to this, since on all levels except the highest, they represent individual ships.

paper—on a coated sheet the ink floats on top of the paper and even though it may be varnished is more subject to removal by abrasion. Try this test: take an ordinary pencil eraser and stroke, with even pressure in one direction, on the surface of a glossy counter sheet. Count the number of strokes it takes to remove the color to a noticeable degree. Using the same eraser, do the same thing to a matte finish counter sheet produced by SPI. You'll notice that it takes *fewer* strokes of the eraser to degrade the glossy surface! To make sure the experiment is fair, try to do it on similar ink colors of similar density. If you don't want to ruin your counters, use part of the frame of the counter sheet to make the test. Incidentally, if you simply *must* have glossy counters, the best glossy counters you can get are SPI counters sprayed with acrylic plastic. Buy yourself a can of Krylon Crystal Clear or its equivalent, and before you punch out the counters spray them lightly several times (in a well-ventilated room), allowing them to dry between each coat. Doing this gives you a high gloss finish *plus* the ink is bonded to the uncoated paper under the spray coating, making for the most durable set of counters one could ask for.

The counters in almost all games have historical designations on them. Most of the time these designations are printed in a type size so small as to be readable only with close inspection. Because of the space problem on counters this is inevitably so—to a greater or lesser degree. In most games this is not much of a concern because the designations are irrelevant to the play of the game. More and more, however, game designers are producing games that employ historical command structures and higher unit organization in more realistic ways. This requires that the player actually *use* the designations during play. This is a burden both on the graphic artist and on the player. What I would prefer to do is to substitute a number code for the historical designator.

A well designed code system would have several advantages over using the historical designator: 1) It would be only four digits long (in most cases) and could therefore be bigger and more readable. 2) It would be more logical and easier to interpret than the sometimes less-than-straightforward military systems. 3) It could be related to an order of battle list that would give the corresponding actual historical designations. Using such a system would allow players to have the best of both worlds: they would be employing their counters in an historical manner but the historical relationship would be easy to grasp and large enough to see at normal playing distances. Take for example, a game like *Terrible Swift Sword* (a tactical level treatment of the Battle of Gettysburg). In the actual game there are two lines of type on every counter that give the historical designation of the unit. This gives you the regiment/brigade/division/corps of the unit. All this information is necessary for the play of the game, but because it is so lengthy in historical form, the designation is set in very small type—its effective size is about 6 points, i.e, only about two-thirds the size it should be. Using the system I propose, one could represent the relationship in a single four digit number that could be almost as large as the primary value numbers on the counters. The order of presentation would be reversed to read corps/division/brigade/regiment. For example "4312" would be the second regiment of the fourth corps, third division, first brigade. A unit representing the brigade commander would be numbered "431" and a unit representing the division commander would be numbered "43". In this way the hierarchal relationship is clearly and unequivocally expressed in quite a simple and readable manner. None of these numbers would be the *actual* historical division or brigade numbers but they would nevertheless be historically accurate when related to a key listing provided in the rules (so that players could see the historical names of the units if they desired). This system may never see the light of day because the typical designer is very territorial when it comes to giving up historical "flavor" for rationality—

and a vocal minority of players would scream if the 118th Bathwater Light Hussars didn't have their glorious name blazoned on their unit counter. So much for common sense.

The Rules

At SPI the art department has more to do with the rules than one might logically expect. This is because I am something of a fanatic about rules systems and proper explanation and organization. Many of the basic formats used in SPI rules are traceable to standards and procedures set up by my staff and me. SPI's rules are distinctly an SPI product—the typical set of non-SPI rules are rambling, conversationally written and weak on organization. Some people like such rules—I don't. I'm supported in my feelings by most gamers. A recent survey showed that SPI rules provide better and more accessible information on how the game is meant to be played. The same survey showed that although SPI's legalistic rules style may draw some cheapshots from those that are not up to such rigorous organization, the average gamer *wants* the rules to be precise, exhaustively detailed, and unambiguous. Let's face it: rules are not exactly light reading—the number of concepts and procedures to be explained in detail can hardly be dealt with in a few easy paragraphs of colloquial English. The closest analog to a set of rules would be a set of computer program statements Such comparisons inevitably draw knee-jerk negative reactions from certain people who equate technology and organization with dehumanization and sterility. What such people fail to see is that such systems are *enabling* devices—and are not the thing itself—the thing is the actual play of the game. The rules are means to an end—and they must be highly organized and efficient means to serve the complexity of wargame play.

One of the minor criticisms I've heard about SPI rules is that there are few light touches of humor to be found in them. To my mind this is all to the good. Rules are *documents*—to be referred to time and again. Have you ever read a joke more than once and laughed? And to be quite frank, most of the people (at SPI and elsewhere) that write rules are not what you would call great sophisticated wits. The high-school level humor that sometimes finds its way into their rules writing is nothing to be savored. Rules writing is inescapably *technical* writing—not literature. It's object is unequivocal communication—not entertainment. The entertaining part is supposed to be the play of the game.

The game designer usually starts with a very rough outline of the rules to the game. The developer has the primary responsibility of turning this into a set of complete rules statements. This is an extremely difficult thing to do well. How difficult it is to do can only be appreciated by those that have tried it under professional conditions. With all the testing, proof-reading, and evaluation that SPI rules undergo in their life history, there are few extant rules booklets that don't still carry mistakes, inconsistencies, and omissions. SPI's policy of issuing errata after their rules are published has brought criticism from some quarters—particularly invidious are the thinly veiled gibes of competing companies that do not *admit* their mistakes and continue to make their customers suffer with error ridden rules until they get around to publishing a second edition of the game. One must be honest about the limitations of the rules generation process—to create *flawless* rules on the first go round is virtually impossible unless the game is so simple as to be irrelevant. Beyond simple typos and plain oversight, there will always be the possibility of alternate interpretation of given statements—because the player is *not* a computer: he's a thinking human who brings his own background and mind-set to the reading of the rules.

When the game is turned in to the Art Department, the copy editor and the graphic designer go over the rules for styling and formatting, plus they attempt to make sure that the rules are complete and comprehensible. Usually a publisher will develop a "house style" in which all their rules are set. The graphic artist must see to it that this style is as flexible and as effective as possible.

The style that I've set-up for SPI is one that incorporates the following main features:

1. Rules are to be indexed according to Main Section and Sub-section.
2. The body of the rules shall consist of main sections introduced by an overall statement (called a General Rule) and divided into subsections (called Cases). These sections shall all be numbered using a decimal numbering system that expresses the hierarchy of the rules.
3. Whenever possible, the ordering of the rules shall follow the sequence of play.
4. Sections and subsections shall carry concise titles expressive of their content.
5. Rules shall be set no smaller than 8 point type and printed in booklet or folder format.
6. Separate sheets will be kept to a minimum.

The graphic designer should gain enough familiarity with the rules of the game so that he can adequately judge the need for "graphic intervention" to relieve the player of needless work and also to insure that any procedures in the rules are consistent with his graphic solutions and vice-versa. He should insure that sufficient graphic and written examples are included in the rules—many times a rules writer will be guilty of making too many assumptions about what the gamer "already knows".

Final copyediting of the rules, at SPI, is the responsibility of the Art Department. This usually involves a lot of consultation with the game developer/rules writer in order to ascertain exactly what he *meant* when he wrote what's on the manuscript. Once they are marked up, the rules go to the typesetting machine operator. The operator keyboards the copy using a fairly sophisticated phototypesetting device (at SPI) that transforms the typewritten manuscript into strips of justified type on photographic paper. These strips (called galleys) are read for errors and then turned over to the layout artist who will cement them onto page forms. This *paste-up* of the rules is then xeroxed and read again (by both the developer and the copy editor) in order to catch any errors that got by the first time. Corrections are re-set and pasted over the erroneous portions of the paste-up forms. Most gamers don't realize that what they see as a continuous page of orderly type is actually the result of dozens of pieces of paper precisely pasted into position by a human being. The paste-up forms are then *imposed* on an eight-page master form in the order that will produce a properly sequenced set of pages when folded and cut by the binder. A same-size film negative is made of this page form and given to the printer who makes a proof copy of the negative for final checking by the publisher. Corrections at this stage are fairly expensive and tedious to do—so only really vital changes are made. When corrected for the last time, the negative is "burned" onto a metal plate and printed on a sheet of paper which is ultimately collated, folded and cut to produce a finished rules booklet. Once the first edition of the rules has been sold it will generate a certain amount of errata arising out of the questions asked by the players. An errata sheet will be published in the same format as the rules. Whenever possible the second edition of the rules will be corrected to incorporate this errata.

My favorite fantasy (regarding rules) is to have a master file of hex-grid wargame rules that would cover every possible situation that could occur in a game. These master cases would be precisely and lucidly written and organized into a data retrieval/word processing system so that entire blocks of rules could be called up electronically by keying in a string of code numbers. The developer would then add whatever minimal necessary names and dates and the whole body of rules would be automatically typeset. Every case would have a master reference number and a computer program would make sure that every case number that needed cross-indexing what get it. I'm actually pursuing this fantasy to some degree—I've started to write trial master cases onto the diskettes of SPI's typesetting device (which is a small computer itself). Any reader familiar with word processing and computer text-editing, however, will be aware of the tall order my fantasy makes. I estimate that it would take about six man months to develop a workable basis for such a complexly interrelated group of statements and then about a man-year to actually write and de-bug the statements. Even this may be optimistic, since I've only nibbled at the edges of the problem.

It would be a boon to editors and gamers alike to have such a system working for them. The clarity and preciseness of the rules would take a quantum leap forward and the flexibility of development in game systems would increase mightily. Game testing could proceed with more finished sets of rules. Annoying minor typos could be forever banished. Laborious typesetting tasks and long production times could be reduced. Ah, the millenium would arrive for rules compulsives such as I.

The Playing Aids and Tables

When I first began designing wargames, not much had been done in the way of systematizing the interface between the rules and the actual play of the game. Before I became professionally involved, one of my hobbies was to take an existing game and build a graphic system for it. Eventually, as you might guess, this hobby became a business—and here I am. I've always considered myself to be more technologically oriented than most artists and I suppose this inclination shows in my work. Games are after all, paper machines. With proper engineering they should reduce the amount of work that the player must perform. Deciding how much to support the play of the game must be based upon the complexity of the game, benefit of the system, the effort required to execute the system and its commercial feasibility.

Although sometimes a game designer or developer will be able to create a playing aid of value, more often than not they must be designed by the graphic artist who appreciates the problems of the player. This can sometimes cause a problem in the game design process—the graphic designer may create a system that replaces one designed by the developer and thereby cause a certain amount of rules juggling to be performed.

Over the past several years, I've developed a number of solutions to the interface/information problem. Some of the basic ones are described in the following list:

1. *The Integrated Turn Record Track.* This is a track of game-turns with all the timed activities signalled on it. The player moves a game-turn marker to record the passage of time and when it is placed in a game-turn space, that space reads out information necessary for him to know at that time. This information can be any one or combination of the following: reinforcements or withdrawals, weather, supply levels, victory consideration, scenario start and stop points, political events, production considerations, invasion windows, and just about any action required to be performed on a specific game-turn. The advantage of such a track is that it concentrates all of the information a player needs and coordinates it with *when* he needs to know it. This allows the player to "forget" all of this detail until the time comes for him to deal with it.

2. *The Integrated Production Spiral (or Track).* This relative of the Turn Record Track enables the player to plan and forecast his unit production. The track is really a combination of two tracks, one of which parallels the other and shows the player the arrival time of any new unit once he knows the time of initiation of production.

3. *The Phase Record Track.* In very complicated games, there may be many phases and segments to the turn. Incorporating the sequence of play into a track allows the player to record where he is in the game-turn while at the same time reminding himself what he's supposed to be doing.

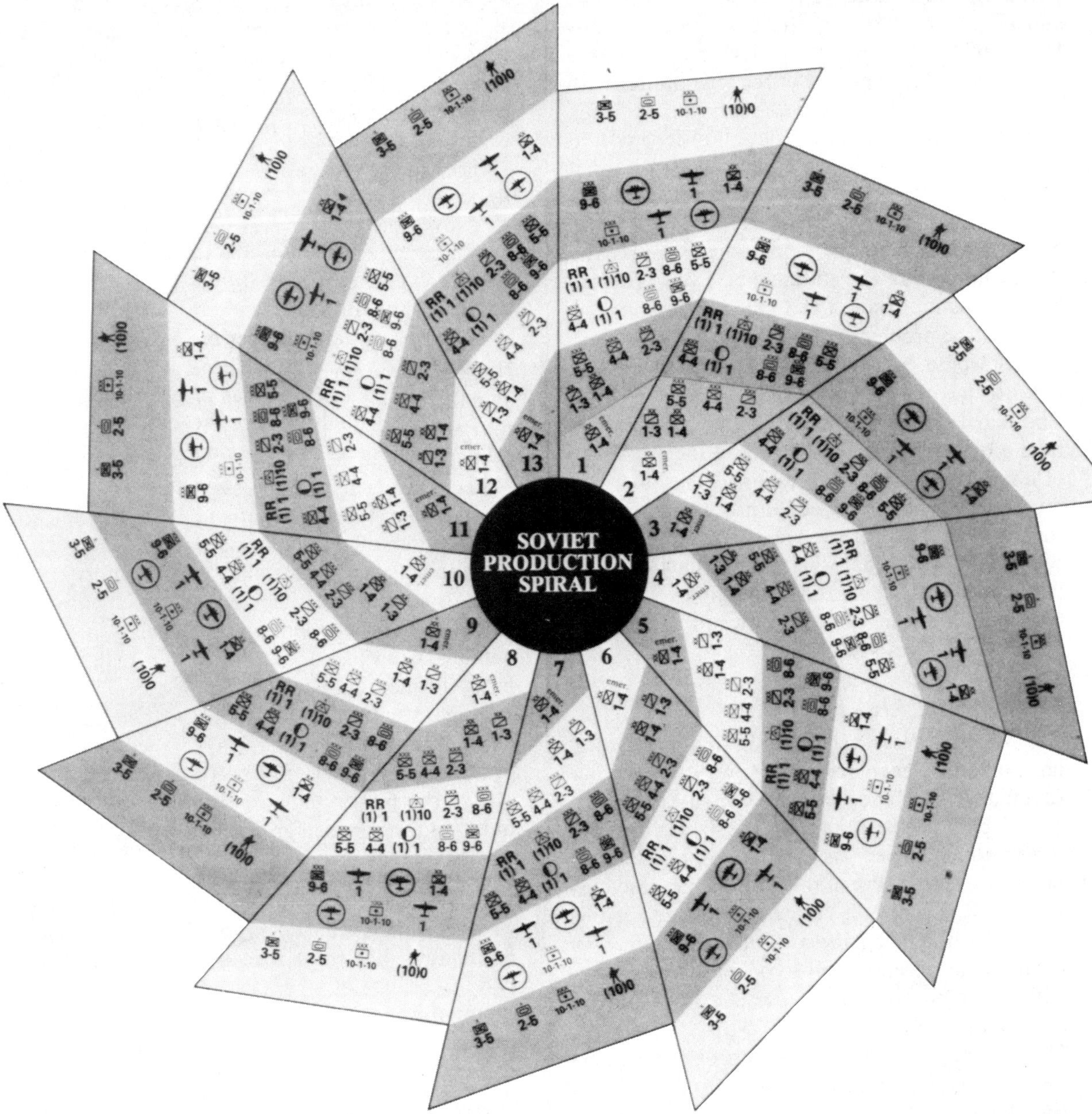

Pictured above is perhaps the most sophisticated paper playing aid yet designed at SPI — the Production Spiral from the *War in Europe* game system. This was a direct development from the linear, *Global War* Production Track. The spiral form solved the problem of accounting for production over a great number of Game-Turns. Notice how all the units indicate their production times by their positions on the spiral — the arms of the spiral emanate from the current Game-Turn; the spokes of the spiral line up with future Game-Turns (i.e., the time at which the unit in production appears in the game).

4. *The Point Record*. This is nothing more than a paper abacus used to record an ever-changing score of points accumulated or lost. Its better than simply writing such points on paper because its faster, easier to access and can have information built into specific points on the track. Tracks such as these are used to record Victory Points, Supply Points, Air Points, Political Points, etc, They require the allocation of a certain number of counters as register markers.

5. *The Status Display*. This is used in many games to record the information surrounding a given unit, or units, engaged in some complex activity which cannot be wholly represented by the playing piece and the map system. In air games, such a display records speed, armament, climbing and diving, or gives mission information etc. In a naval game (for example *Fast Carriers*) it could record the composition of Task Forces, show the condition of aircraft on carriers, give fuel readouts, etc. In a land game, the display might use a system of tracks and holding areas to indicate the composition of some higher organization, the organic supply capability, the morale level of a force, etc. Status displays are usually a system of interrelated tracks and holding areas that require a number of record markers and/or the actual playing pieces themselves to be moved or placed on the display.

6. *The Integrated Combat Results Table*. This is an otherwise conventional combat results table that has its headings and columns so arranged as to automatically account for the effect of terrain, supply, or unit type on combat. It most often slips the ratio or differential headings in order to make the same results column applicable to many situations depending upon the change in some variable such as terrain or supply state.

7. *The Variable Set-Up Key*. This is a set of unit pictures printed on a separate sheet or directly on the map. Each picture has associated with it the key letter or number of a scenario or scenarios. In this manner a number of different set-ups can be collapsed into one. When printing the set up directly on the map, a color change is often used to represent a scenario change. A variant of this key is the Formation Display (developed for *Highway to the Reich*) that shows the player the proper organization of the counters and also serves as a set-up and reinforcement staging area keyed to the Turn Record Track and the map.

8. *The Compartmented Game Box*. This is simply a vacuum-formed plastic tray that gives the player an adequate number of storage compartments to sort his counters into for future ease of set-up. It's a totally obvious piece of equipment that no one thought to give the gamer until SPI developed a packaging system for its games.

9. *Numbered Hex Grid*. This is a regular system of four digit numbers printed directly on the map hexes to unequivocally identify them. Once again, an obvious idea that no one implemented until SPI came along.

10. *Indexed Case-Structure Rules*. A format for rules construction that organizes the rules into numbered sections each of which is introduced by a general statement and then followed by short detailed statements, decimally numbered in relation to the section number. This General Rule/Case structure and its attendant numbering enables a player to more easily find specific rules that deal with the specific question he has in mind.

When a designer attempts to aid the player by providing him with a graphic device, of any sort, he must be careful that the neat little system he comes up with doesn't actually *add* complexity to the game system. Things to be avoided are: excessive use of abbreviations; too many markers operating on a single register (sometimes a pencil and paper *is* better); systems that are so cramped by lack of space that they become difficult to use; systems that are *larger* than the playing map or that take longer to set up than the game itself; and any system that takes longer to operate than the maneuver portion of the game-turn.

There is no easy formula for developing graphic systems that aid play. Most of the really good ones are stunningly obvious—once you see them in operation. Much of the success one will have will depend upon being able to project oneself into the position of the player who will have to deal with the finished game. Whenever possible, the graphic designer should actually play the final version of the game using the test components. Unfortunately, this is sometimes difficult to do since games take a lot of time to learn and play—and the artist *doesn't have* a lot of time in a commercial environment.

HOMEMADE GAME COMPONENTS

Just beneath the surface of every wargame *player* lies a wargame *designer*. Most gamers have had the impulse to either create their own games or to add something to an existing game. One of the major stumbling blocks to home-brewed games is the creation of components—wargames are "art-intensive" projects—and most gamers are unskilled and unversed in graphics. I can't of course, make artists out of all of you by writing a few thousand words in this book. What I can do is give you some idea of the basic materials, equipment, and techniques available—and some suggestions on how to go about designing and creating your own components.

The first thing to do is to get out your entire wargame collection and look at each game with a question in your mind: why did the artist handle the components the way he did? You must raise your consciousness and increase your sensitivity to graphics. The typical non-artist does not consciously *see* what he looks at—he merely reacts to it. This is not meant as a slur on anyone's perceptions—it's simply a fact that the artist's viewpoint is a special one and the would-be designer must emulate this state of mind if he is to have any hope of his components rising above the Crayola level of appearance.

Some of what we call artistic ability is inherited—but much of it is learned and/or "absorbed". Whether or not you have any intrinsic ability, you should be confident that you can develop your sensitivities enough to produce satisfactory components. And, in the bargain, you may gain some insight into the problems of graphic design and execution. Good design sense is a valuable acquisition even if you only dabble in the actual production of components. Good *technique* is necessary to produce high quality components. There are actually two sets of overlapping techniques. One set of techniques is necessary for creating component art that is to be printed (i.e., art for reproduction) and another set for producing a hand-crafted, one-copy-only pieces of art. It is this latter set of techniques that the wargamer will find the most use for since few of you are going to go into wargame business. I'll concentrate on these *direct* techniques and merely touch upon the techniques of art for reproduction.

Components for a single game can be produced directly for only a few dollars or you can spend over a hundred dollars. The cost depends on how highly finished you want your components to be and how much you're willing to spend on materials and services to achieve that degree of finish. A skilled craftsman can produce a set of components that are indistinguishable from professionally printed art—in fact, in some respects they will be superior. Achieving such a level of craft requires practice and patience and a broad familiarity with the tools and materials of graphic art. As with any technical skill, graphic production seems like magic to the uninitiated. Of course, it isn't magic—it's a lot of hard work. The best equipment in the world won't do the work for you—but having some good basic tools and materials is a great asset. If you're on a budget, buy only the most basic equipment, but don't go third class—art equipment quality varies directly with price. Since the price of even the low-grade equipment is not inconsiderable, you may as well "bite-the-bullet" and pay the price for the higher quality equipment. Good art tools will last a lifetime and have many applications to non-art purposes.

What follows is a list of art equipment and materials that are used in direct production of game components. Those that I consider *necessary* to all but the most rudimentary graphic production are marked with an asterisk (*). The other equipment listed are nice-to-have items.

Graphic Supplies

TOOLS AND EQUIPMENT:

1. T-square, stainless steel, 36" length. (one)*
2. Triangle, plastic, 60°, 14" length. (one)*
3. Triangle, plastic, 45°, 12" length. (one)
4. Triangle, plastic, 60°, 8" or 10" length (one)
5. Drawing template, plastic, "Geometric Designer" by Timely Products (or equivalent). (one)*
5. Knife, "X-acto" nr. 1 handle/nr.11 blades. (one)*
6. Knife, utility and blades. (one)*
7. Ruler, stainless steel, 18" english/metric, "Gaebel" nr. 624 or equivalent. (one)*
8. Technical fountain pen, "Castell, TG Series" nr. 0, or equivalent. (one)
9. Giant bow combination compass, 10". (one)
10. Drawing Table or board, 42" or larger. (one)
11. Light box or table, 36" or larger. (one)
12. Electronic slide rule calculator. (one)
13. Tape dispensers, 1" core and 3" core. (one each)
14. Single-edge industrial razor blades. (one package)
15. Kneaded eraser. (one)*

INKS, DYES, PAINTS, SPRAYS, AND MARKERS

1. India ink, "Higgins Black Magic". (one bottle)*
2. India ink, "Artone Extra Dense". (one bottle)
3. "Dr. Ph. Martin's Bleed-proof White" water color. (one bottle)
4. "Dr. Ph. Martin's Radiant Concentrated Watercolors" aniline dyes, Set B or individual colors. (as needed)
5. "Winsor Newton" designers gouache (opaque watercolor). (in tubes, as needed).
6. Clear acrylic spray, "Krylon" or equivalent. (one can)*
7. Clear matte acrylic spray, "Krylon" or equivalent. (one can)
8. "Letraset" matte spray. (one can)
9. Fine point marker pen "Pilot Razor Point", black, red, and green. (two each)*
10. Fine point marker pen "Flair", black, turquoise, brown, red, green. (as needed)
11. "Pantone" Color markers nrs. 298, 466, 154, 375, 102, 422 and others as needed. (one each)*

PENCILS, BRUSHES, AND APPLICATORS

1. Pencil, "Mongol" light blue nr. 845 (two)*
2. Pencil, "Venus" graphite nr. 3H. (two)*
3. Pencil, "General" charcoal white, nr. 558
4. Brush, red sabel watercolor, nrs. 000 through 2. (as needed)*
5. Brush, nylon, 2" flat. (one)*
6. Burnisher, fine point, for transfer lettering. (one)
7. Bottle and applicator, 16 oz. for rubber cement. (one)
8. "Valvespout" thinner dispenser. (one)
9. Rubber cement pick-up (remover). (one)

TAPES AND ADHESIVES

1. "Elmer's Glue-All" white glue or equivalent. (one bottle)*
2. "Best-Test" rubber cement, quart. (one)*
3. "Bestine" rubber cement thinner, quart. (one)*
4. "Duro" contact cement, can. (one)

5. "Duro" contact cement thinner, can (one).
6. "Best-Test" one-coat rubber cement, quart. (one)
7. Masking Tape, 3/4"–60 yds. (one roll)*
8. "Magic" Transparent Tape, nr. 810, _". (one roll)*
9. "Chartpak" or "Zipaline" charting tapes. See catalogs.

PAPERS AND BOARDS
1. Ledger bond, 11" x 14" pad. (one)*
2. Tracing paper, 11" x 14" pad. (one)*
3. Tracing paper, roll, 36" wide. (one)
4. Cross-section graph paper 17" x 22" pad, 8 or 10 lines per inch. (one)*
5. "SPI" numbered hex-sheets, 23" x 35". (six or more)*
6. "SPI" blank counter sheets, Type A or B. (six or more)
7. "Bainbridge" nr. 172 illustration board. (as needed)
8. "Bainbridge" nr. 24 smooth mat 20" x 30" board, (as needed)

LETTERING MATERIALS
1. "Prestype" Dry Transfer Lettering or—
2. "Letraset" or—
3. "Paratipe" or—
4. "Pressure Graphics" (available by mail only) 1725 Armitage Court, Addison, Il 60101

STORES AND SERVICES TO BE USED
1. Office copier (Xerox or equivalent)
2. Short-run Printing Service
3. Photo-stat shop
4. Art Supply store
5. Hobby shop
6. Large stationer's
7. Hardware/building supply store

Before you actually buy any equipment, visit your local art materials store (the bigger the store, the better). If the store has a catalog (many of them do) see if you can get a copy. Art store catalogs are an education in themselves. There are all sorts of great little goodies not mentioned in the above lists that you may have use for. Some of the items may be available in large stationery stores—but you'll usually pay a higher price for them. New York City has many very good art supply stores. Look in the Yellow Pages commercial phone directory for their names. Almost all large metropolitan centers have at least one good art store—find it.

It also might help if you have any friends who are professional artists. For the price of a couple of beers you may be able to get a lot of valuable information. You may also be able to arrange borrowing some of the more expensive items of equipment. If your artist friend is also a wargamer, he's probably already involved in designing his own gaming equipment. See if you can pool your resources and make a joint venture of it, thereby gaining the advantage of more than one viewpoint.

General Advice

Take your time. Don't rush out and buy a lot of equipment and immediately begin to bash away at making your own components—it's almost guaranteed to get you into trouble and perhaps discourage you permanently. Try a small project first and only after you've familiarized yourself with its requirements. Visit a good library or bookstore and try to locate a few books on graphic art techniques. A particularly good one to get is *Commercial Artist's Handbook*, by John Snyder, published in 1973 by Watson-Guptill, New York.

Before you attempt to do the final version of your project, make several sketches or roughs. If you're doing counters, do trial versions of each type before doing the whole counter mix. For maps, do several small samples of terrain that includes all the various features you intend to include. Show your sketches to another wargamer (even if he's not an artist) and discuss what you're trying to do. Think about the sequence in which you're going to do the work—a lot of time can be saved by properly planning out a job before you begin.

Making a Game Map

This is the most extensive project one can attempt and the one most often tried. Amateur mapmakers are very often dissatisfied with the results they get and are not quite sure why. Keeping in mind the description of professional game-mapmaking, you should work with a good source map, scaled to the proper size, and use a properly thought-out scheme for the terrain. The biggest mistake made by amateurs is being too heavy-handed in the application of color and line. Everything looks too bright, too raw, too boldly drawn on most amateur maps. Since all the elements on the map are shouting for visual attention, the result is chaos. When working directly it's important to remember that it's almost impossible to make anything *lighter* once it's drawn, but it's very easy to make things *darker*. What this means is that you should *build up* the features on the map by stages rather than attempting to get it all done in one shot. If you have several Pantone markers available, use the lightest versions of the colors first. It's always possible to intensify the color by going over the same area.

When using Pantone markers, the colors should be laid down in smooth, even strokes in the following order:

Yellows
Tans
Grays
Blues
Greens
Red
Dark Browns
Black

As you can see, you're basically working from light to dark—and this is done so that the least amount of *poisoning* occurs. With markers, working light colors after strong colors often draws some of the strong color into the light color *poisoning* the color field and the felt tip of the lighter color marker. When you work the colors in the proper order (though it may seem strange at first) you also have better control over the general intensity of the map, and you should be able to avoid the problem of the "loud" map.

You will notice that felt markers *bleed,* i.e., spread, into the paper surface. This is unavoidable but it is controllable. Do a few practice strokes on the kind of map sheet you're using and watch to see how much the color spreads. Allow for this spread when doing your final map. Notice that different colors or types of markers will spread at different rates. This bleeding effect can sometimes be used to advantage to soften the edges of terrain. If a color goes on too heavily, you may be able to lighten it somewhat with a cotton ball soaked in rubber cement thinner or by swabbing on laundry bleach. Do a test before trying this—and be very careful when using either color remover: one is flammable and the other is toxic. When using markers be sure the room is well ventilated—the markers are very volatile and their fumes should not be inhaled. You should also cap the markers as soon as you finish with them (even mid-job) since they dry out very easily and become useless.

Sometimes markers will bleed right through the hex sheet and discolor whatever's underneath—so don't work over a surface that can't stand being stained. If you're tracing a valuable map on a light box, interpose a sheet of clear plastic between the hex sheet and the map. The plastic can always be wiped clean with a swab of thinner or rubbing alcohol.

Some effects that are possible in printing are not easily accomplished in a hand made map. The color transparency of printing inks is difficult to duplicate *cleanly* in markers. The use of patterns is also restricted in handmade maps. The adhesive-backed, pre-printed sheets sold in art supply shops are not really suitable for use on a surface that will be played on—although they're fine if your map is going to be printed. In any event, most of these patterns are available only in black (the color separation and printing process renders them into the colors you see on professional maps). Even those patterns available in dry transfer form are not good choices—they too will ablate from the surface unless heavily sprayed with acrylic. If you want textures and patterns, you're better off *drawing* them. This can be done relatively easily by stippling with a brush in watercolor, using a stamp made out of a raw potato, sketching with a fine point marker over a solid field of a different color, or by washing droplets of thinner over the marker colors. Always do a trial before using any of these techniques. Variations on these methods will suggest themselves to you as you gain a practical understanding of the characteristics of markers and watercolors.

If you intend to do any of the line work on your map in ink, do it last. This holds true for designer's watercolors as well. Be aware that if you paint large areas with a water-based medium such as ink or tempera, the paper may well buckle and warp. Some of this can be alleviated by painting the back of the paper with clean water, and taping down the map on a smooth surface then allowing it to stretch itself flat as it dries. Generally speaking, you should avoid laying down large areas of opaque color on a game map.

Lettering on maps can be handled simply by using very neat hand-printing or one can invest in some sheets of transfer lettering. Unless you're very good at it, don't try to be fancy with your hand lettering. Draw some very light pencil guidelines on your map and print cleanly. Use ink rather than markers. It wouldn't hurt to practice your printing on some graph paper—after printing a couple of dozen words you'll start to recapture some of the penmanship you once had.

Mapsheets can be mounted onto large sheets of cardboard (or wood for that matter). A semi-permanent mount can be done using rubber cement—coat each surface evenly with cement and let it dry thoroughly. Place two large sheets of tracing paper on the mounting surface, overlapping slightly in the center. Place the mapsheet on top of this and smooth it out. Now lift up one side and pull one of the tracing sheets out by a few inches. Smooth the center of the map with the edge of your hand and the glue will grab. Always smooth from the center outwards. Slowly withdraw the tracing sheet, smoothing the map into contact with the base as you go. Now do the other half the same way. Use a rubber cement pick-up to clean away any excess dried cement. Seal the edges with a continuous application of Magic tape, wrapping it around to the back of the board. This method of mounting is called *dry mount slip-sheeting.* The same technique can be used with Contact cement for a more permanent bond, or a bond to wood. Mapsheets can also be mounted using the white glue technique described for counter sheets—although this requires a lot of skill.

Making Counters

If you use SPI blank counter sheets, your task is somewhat simple. All you have to do is to decide whether to hand-letter your values or to use dry-transfer lettering. Whichever way you choose, do it *before* you punch out the counters from the sheet. If you're going to hand letter the numbers and designations, you'll find that not all felt-tipped pens work over the printed color fields of the counters. This is because of the wax in the ink. Pantone makes its markers in a fine-tipped version that is suitable for relatively small lettering—but for really tiny lettering such as designations, you're going to have to use india ink and a steel pen. A technical fountain pen can be used if held properly (almost perpendicular at all times) but this will be slow.

Dry transfer lettering makes for professional looking counters. With a little practice anyone can produce neat expert lettering. The drawbacks are: expense, tediousness, and the need to protect the counters with a careful spraying of acrylic or lettering fixative. The cheapest source of dry transfer lettering I've found is Pressure Graphics in Illinois. They're a mail-order house specializing in transfer materials and markers (not Pantone, but "Admarkers" which are just as good). Write them for a catalog and a sample sheet (you'll get this if you can legitimately use your company letterhead). Their lettering is not quite as sharp as Letraset, for example, but then in this application sharpness is not super-critical. All the makers of dry-transfer materials have their own catalogs which are useful sources of information on technique and applications. You'll also see in them many related materials not discussed here. Most of the companies make sheets of just numbers—that which you'll have the most need of in doing counters. Use as your guide, the discussion of type in the passages on professional counter design. You'll find the Helvetica Medium styles are the most obtainable number sheets. As with any of these materials, you should do some practice counters before proceeding with the full scale project.

The best counters are those that are made completely from scratch. They are also the most work. If you have access to a photostat service, you can do your counters larger than final size (on nonreproducible blue graph paper) and have them photo-reduced. This also allows you to do only a few master counters and get as many copies as you need photostatically. A really beautiful set of counters can be made using photostat images with aniline dyes washed over them for color. This requires that the stats be "washed well" to remove any hypo residue. You then place the stat on a scrap board taping the edges all around. Use a cottonball dipped in clean water to evenly moisten (not soak) the stat overall. Dilute the aniline dye in clean water (one part dye to two or three parts water) and use a damp cottonball dipped in the dye to apply it to the surface of the stat. Apply the dye in continuous circular motions. The color should be weaker than that desired so that it may be smoothly built up by repeated applications of dye. This is a messy process but it results in the best looking counters you'll ever see. Let the dyed stat dry thoroughly then buff it with a dry cottonball to remove any excess dye.

Now mount the stat on a piece of mat board and cut out the counters using a utility knife or an X-acto. Mount the counters using white glue cut with water (two parts glue/one or two parts water). Paint the glue on the back of the stat using a nylon brush. Paint the mat board with a thin coat of glue. Smooth the moist stat onto the mat board. Place a sheet of tracing paper over the stat and weight it with heavy books. Let it dry through. When cutting out the counters, use a steel T-square as a guide edge. Always cut *away* from your non-blade hand. Make the cuts using repeated medium strength strokes. Never try to cut through in one heavy stroke—you'll either ruin the job or cut yourself. Hold the knife vertically when cutting. Have a piece of scrap board under the counter sheet to prevent the blade from getting dull or cutting into your table. Tape the work down securely when cutting. Don't rush. Don't cut all the way to the outer edge of the sheet—leave a dead frame around the counters to hold them in place. Make all the cuts in one direction first.

An alternative to using stats is to make xerox copies of your counter master on colored paper. Most markers will smear xerography and dyes don't take well on the sulphite paper used in copying machines. Some short run printers work cheaply enough so that you might want to try using them to print your unmounted counter sheets. This is only worth it if you have a *lot* of counters in your game. Another possible approach is silk-screening counter images directly onto SPI blank counter sheets. There are many books available on silk-screen technique and the process is easy to master (for the level of printing we're considering).

If you're going to make your own counters at all, take the time to do a good job. Looking at illegible counters while playing a complex game can be a real impediment.

Presenting Rules

In a later section of this book I'll detail the style that we use at SPI. Here, however, a word about how rules look is advisable. If you expect your game to be played by your friends and gaming associates, it's worth the effort to do a good job of presentation. Needless to say, you should write an outline, then a first draft (at least) before writing your finished rules. They should be neatly typed. If you've got a carbon ribbon this will help. Type your rules on 8½ x 11 smooth finish bond. Type only 25 lines to the page and make each line only about 50 characters long if the typewriting is to be the final form of the rules. If your typewriter can manage it, set the line space on "1½" (if however, you're typing a *manuscript* that is to be set into *type*, always double space). If you're going to print your rules, photo reduce them to about two-thirds of the original size of the typewriting. This cuts down on the number of sheets and gives your rules a more "typeset" look. Leave generous margins and inter-section spaces—typewriting is always a little harder to read than typesetting. Bind your rules in a report folder (the kind with a spring clip) so that revisions can easily be inserted. Charts and tables should be typed on separate sheets. The T-square and triangle can be used to rule in lines with a technical pen.

Packaging Your Game

You may wish to use an SPI plastic box to package your finished game—you can even make your own cover sheet. Presentation is important if you want your friends to take your game seriously. Alternatively, you can construct your own custom designed trays and boxes out of mat board finished with acrylic spray paint. The skills required here are similar to those used in model building. The construction of custom designed packages can become a hobby within a hobby. Many valuable play aids can be built into such packages that could never be done commercially. Compartments, racks, storage by order of appearance, range templates of clear plastic, "moveable" terrain, and magnetic counters are only some of the possible things that can be constructed once one gains some facility with art materials. Even if you have no interest in creating your own games, many of the games in your collection can be enhanced by the addition of custom made components.

SECTION IV

Notes on Game Design

by Stephen B. Patrick

There are numerous steps involved in designing a game. Often they are not recognized as separate steps, but experience has shown that when these various stages are broken out into their separate elements, it is easier to identify the problems each one poses. It is also easier to progress through the design of a game in a methodical fashion. For the amateur game designer, and even for the experienced designer, it is important to recognize these steps in order to do a complete job of designing a game.

Taken in the order that a designer has to deal with them, research is the first step encountered in game design. A person performing the research needs a knowledge of the specialized sources and the relationship between research and the other functions of design and development. It is important to know things as basic as how to find a book on a specific subject. It does a designer no good to sit next to the biggest and best military history library in the world if he does not know how to make use of it.

The next stage in design is Quantification. A simulation consists of numbers, and a designer does not simply pull a number out of the air and say, "We'll try 4 for movement and see what happens." Quantification relies on base values. In *Jutland,* for example, the base value was the German eleven inch, 40 calibre gun. Everything else was calculated based on that one gun. In *1914,* the base was the German *Landwehr* brigade. That unit received a value of 1 and everything else became an analog of that unit. In picking the base value, the designer looks for the most common unit, because the most common unit is the one that will have the least error in its strength evaluation.

The relationship between design and development and other functions is critical to the Integration phase. It is at this time that the designer alternates between playing the game and continuing to research it. If a designer is doing anything "creative," integration is the moment of truth—the moment when he actually brings the game to life. Once a designer has the game working well enough to achieve integration, he has a going game. It is very difficult to explain how to do this. A good technique is to modularize the rules, such as setting out separate categories of movement, combat, terrain, and the like. Basically, everything is a function of movement and combat. The trick is to break the elements into as many little blocks as possible—at least into half a dozen. Then the designer can isolate each element and decide, for example, that movement will proceed in a particular manner. This is a problem area in amateur games. In those games, the designer often is not sure of what the movement allowance will be even when he is mid-way through the game.

The next stage is Development. This is a separate phase that SPI pioneered in creating. The development stage provides the developer with an opportunity to try out the various options presented by the designer. The basic function of development is to refine the design. In that regard, SPI uses development as a training vehicle for potential designers. Many people underestimate the amount of work it takes to design games. Anyone who has played *1914* or *Jutland* will see what happens when an historian with a vast amount of information designs a game and omits the development phase.

Finally, there is the introduction of Graphics. Graphics, of course, is the physical presentation of the game. A good designer will have an eye to the graphics even before the time has come to execute them. Some designers regard graphics as little more than an effort toward nice appearance. Companies such as SPI and GDW carry graphics a step further than that by employing graphics as an enhancement of the game's playability. In this view, graphics cannot be relegated to the status of an afterthought. The graphics elements must be considered before the development phase is ended as it can have a substantial impact on the way a game is finally presented.

Once the game is presented for play, the designer must perform a certain amount of analysis if he plans to do more games without repeating the mistakes he has made in the one just completed. Most major companies do this as a matter of course, since it is the only viable means of upgrading quality over a series of games.

This chapter illustrated the SPI approach to these various elements of game design. Certainly the SPI method is not the only feasible approach. It is, however, one of the few approaches that does not depend entirely on the individual personality of one designer. SPI's method is perhaps the best approach for a novice to take if he wants to end up with an acceptable product, even if his only audience will be his friends.

GENERAL CONSIDERATIONS FOR GAME DESIGNERS

A truly simple game will be enjoyed as much by the player who is looking for amusement as by the player who seeks competition. For a designer to reach this degree of appeal, however, he may have to simplify things enormously. If the designer adds extra rules, he will wind up with a bloodbath as far as play is concerned. Therefore, a simple game—one that foregoes these added rules—will ignore the history buff.

In order to assess what the audience will accept, a designer must have some form of audience feedback, because the mix of elements in a given audience is constantly changing, particularly as a company grows. The *feedback* is SPI's name for its ongoing customer survey. In this system, scores range from a low of 1 to the high of 9. A score between 5.8 and 6.3 is the average range for game acceptability. SPI has always consulted its feedback before a game is done. If the feedback is favorable, then the game is done; and if response is unfavorable, the game is not done. SPI's customers have changed greatly in character from the days of *Normandy* and *Barbarossa* (1969–70), and the design of the games has been modified in accordance with the games' audience.

Thus, there are several factors that affect the acceptability and popularity of a game: historicity (the historical flavor of the game), the historical period, the scale of the game, and what can be termed the cumulative learning capacity of the player.

There are tricks that can be played with a game design to make it appear historically accurate. An example is the use of unit designations or the actual names of the units (e.g., 23rd Division, Scots Guards). SPI made a mistake in *Austerlitz* by failing to supply the counters with historical unit designations. *Austerlitz* is basically a simple game that was purposely denuded of many historical elements for the sake of playability. But many players would have let this "antihistorical" character pass if only unit designations had been squeezed onto the counters. The original argument at SPI was, "Why have someone in the art department spend four hours putting down those little pieces of type for unit designations when it makes no difference to the game?" The fact is that it *does* make a difference in how the game is received by the players. The majority of people like and demand unit designations. A designer who makes a game suited to himself can disregard a great deal of this. But a designer who is going to make a game to be played by the public has to take this sort of thing into consideration.

There is undoubtedly a correlation between an historical topic and the popularity of a game. If a given game system is used to produce, for example, five different games on five different periods, the feedback results will vary greatly among the games. The only thing that can account for such differences is the variety of historical periods represented. Currently, modern games receive the most favorable reader response.

Within a given period of time, some topics will have more popular success than others. An example of this phenomenon is in the science fiction area. SPI proposed *Star Soldier,* a tactical man-to-man science fiction game, two or three times with only a lukewarm response in the feedback. Then *StarForce: Alpha Centauri,* a strategic level game, was proposed. It received a very strong response and was ultimately produced as *StarForce.* Of course, the fact that a strategic game was more acceptable than a tactical game in this example does not mean that strategic games are necessarily more popular as a genre than tactical games. *Star Soldier* was feedbacked again a year after its initial rejection, and it received a high response—based on the popularity of *StarForce.*

Tactical level games do quite well for modern or World War II era games—a fact that might be explained by several reasons. World War II or modern tactical games, such as *Mech War '77* or *Tobruk,* are usually complex, but players do not seem to mind the complexity because the game gives them sufficient return in the form of information. *D-Day, Battle of the Bulge, Barbarossa,* or *Guadalcanal*—non-tactical games—do not give much in the way of informational return.

Another curious phenomenon in the wargame market is the popularity of tanks. People are fascinated by them. Although armies consist primarily of infantry, newspapers and the media in general, as well as the people, are drawn to tanks. Consider the 1973 Arab-Israeli War. A great deal was made of the use of tanks. Usually the vehicle pictured in a television broadcast or newspaper photo was not even a tank; it might be a self-propelled artillery piece or even an armored personnel carrier, but the label "tank" was hung on it and, as such, it got a high priority in the news.

Yet another reason from the popularity of tactical games is the appeal of "lower level" units (i.e., platoons or companies rather than divisions or corps). The lower the level, the more a player can identify with the unit depicted. In a divisional game, the players are moving several thousand men around at a time in a relatively impersonal situation. Most of the players who have had military experience, in fact, never took part in division level tactics. But in a man-to-man game, such as *Sniper!,* players identify with the pieces enough to give them names in some cases, and even regard one of the pieces as themselves!

For an example of how these factors are intermixed, examine the game *The East Is Red.* The title was chosen for inclusion in the feedback because modern period games are popular (although the problem with modern games is the shortage of juicy topics due to the fact that these are primarily wars that have never been fought—yet). Because of the shortage of "hard" information on this hypothetical conflict, the designer was able to make the game very simple. No player would hold the designer to the same level of detailed accuracy as would be called for in an historical game, and there simply were not that many details that people would expect him to include. In fact, a game such as this permits the designer to concentrate on making a playable game rather than an accurate simulation, because the only fundamental decision that needs to be made in the course of the game play is who will win and how. There is no historical precedent to adhere to.

When a designer decides what level the game is on—whether simple or complex, strategic or tactical—he must also take into account the availability of necessary information. An experienced designer can often look at a prospective game and determine all of these aspects (although it should be noted that it is not unusual for a proposed "simple" design to take a life of its own and grow more complex, virtually in spite of the designer's intentions).

The designer who is considering a game must also take into account the cumulative learning capacity of the player. The novice player is totally unfamiliar with the concepts that go into a wargame. He is apt to think that a conflict simulation is like chess, until he discovers that he must unlearn chess in order to learn a wargame.

Another problem is that people often misconceive historical situations. A person who has heard of an unusual incident that took place in a particular historical event is likely to question why that incident is excluded from the simulation. Often, such incidents are of such a nature that they would unbalance the entire game. For instance, during the Napoleonic wars, cavalry actually rode out on the ice to attack ships during a particular campaign. Obviously, this is a once-in-a-lifetime situation. It never occurred again, and its inclusion in a game would result in its becoming an important tactic—which would obviously be a mistake. One of the designer's principle tasks is to determine which of the facts he has gleaned from research should be put into the game. Although it would be nice to include all of the historical elements—a task which novice designers often mistakingly attempt—an overabundance of historical elements will result in an unplayable game. Deciding what a game will emphasize is difficult. Does the designer want an exercise in the mechanics of the operation? Does he want to emphasize the logistical aspects? A good designer can "count biscuits" with the best of them, but is that what he wants to do? If the designer does not want the game to stress supply, there is no point in getting bogged down with extensive logistics.

Many people still feel that if an event occurred in history, it should occur in the game. Obviously the designer could put any number of little elements into a game, but he must confine himself to interjecting only those elements that actually add to the game. Whatever the designer is saying in his game, he must be sure that he is saying what he means.

The designer decides how his game will correspond to or deviate from history. He can produce a situation where either side has a chance for total defeat or victory, even when historical precedent denies such a range of outcomes. This is where the designer effectively has the power to play with the players' minds. As long as he has paid homage to the original situation, he is on his own. What he must keep in mind is that the players must still be able to achieve the situation's historical outcome.

Another consideration for the designer is the general situation presented by his subject. A popular and proven approach to a game is to find a situation where one side soundly whips the other for the early turns and then has to hold during the later turns. Such a situation often makes for a successful game as can be seen in *Afrika Korps, Sinai,* and *Drang Nach Osten* among others.

The designer must also consider the popularity of the period with which his game deals. If a previous game on an "unexplored" period has been poorly received, it reflects poorly on the whole period. For instance, there are three or four good games on the ancient era that might have been done if *Fall of Rome*—a poorly done game—had not effectively "poisoned" that era as material for future games. In effect, *Fall of Rome* "cost" SPI four potential games—although *The Conquerors* and *Punic Wars* have recently broken this barrier.

A game can also exert this "ripple" effect on other games by being used as the focus of a series of games. The *Kursk* system, for instance, led to a number of World War II games; and *Napoleon at Waterloo* spawned numerous 19th Century games. An example of a series that never caught on is the aerial warfare series initiated with *Flying Circus*. That game failed to excite much interest—as did the following spin-offs, *Spitfire* and *Foxbat and Phantom*—with the result that several related games, such as *Sturmavik,* were never done.

Perhaps the single most important consideration is that many games suffer in audience reaction because the designer ignored the "essential truths" of the historical situation. The designer must recognize that in a campaign that is considered decisive there are lessons to be learned. If these lessons are ignored, the "essential truths" of the campaign are lost, and this has a direct impact on the game's acceptability. At the same time, designers can get into trouble by coming up with facts that, while true, do not accord with the popular conception of the situation. In some games, the side that won a smashing victory historically does not have an easy time winning the game—perhaps because the historical situation was resolved by a fluke. Of course, in such a situation, the designer must differentiate between an unlikely result that happens by chance and a result that came about due to some major inherent element—such as official policy. For example, the American victory in the Revolutionary War was not a matter of luck: it was the direct result of British policy toward the war. The best British generals and admirals did not want to go to America, and those that did go did not have hearts in it—and, conversely, those who did have their hearts in it were inept. This attitude had a direct bearing on the conduct and outcome of the campaign, and any game on the American Revolutionary War must take it into account. The British attitude was pervasive and must be dealt with as an unalterable "given" in any simulation of that war.

There are other campaigns where the losers lose by way of elementary mistakes that, unlike a national attitude, might reasonably be corrected with the judicious application of hindsight. For instance, the Germans need not have lost at Stalingrad, but they made the mistake of insisting that the 6th Army fight *in* the city rather than permitting it to turn to face the Soviet penetration on its flanks. Had the Germans shifted to meet this threat, the Soviets could have been crushed, as their attack on the German flank was a big gamble to start with. To reflect this Soviet risk, *Turning Point* furnishes the Soviet forces with about five mechanized units and a great deal of infantry, with which the Soviet player is intended to take on the best part of the German army in southern Russia. The smart German player will not repeat the historical error.

The concept of "essential truths" is not limited to the present or past. It applies equally to the future as typified in such games as *NATO* (1980), *Invasion America* (2000), or *StarForce* (distant future). If the designer postulates a future that is not credible, the game will fail. People can be educated on the likelihood of a given future event, just as they came to accept the idea of command control in wargames when they had previously felt that such a problem was nonexistent. Many of the present technological miracles were accurately predicted fifty years ago in a general way. This fact tends to be obscured by all of the more fantastic predictions. If the predictors had better educated the population at that time, many of these seemingly outlandish things which have been developed in recent years would not have been so shocking. A good rule of thumb when dealing with games on the future is that ideas that seem sensible to the designer are more likely to be accepted by the players. The designer need only be sure, in that case, that his idea really is sensible in the final analysis.

Some people balk at the term "simulation" as being pretentious. The difference between an historical game and an historical simulation is that the game usually accords with popular prejudice as to what happened while a simulation tries to show the circumstances under which things really did happen.

The puzzle aspect of many "games" becomes their chief value, because the solution can be elusive. In some cases, it's like a maze. Some games, such as *American Revolution,* are fairly straight-forward. It does not take many games for the British player to realize that his best bet is to wait until the summer of '76 to drop all his units in the Potomac Valley with the goal of reducing the south. On the other hand, a game such as *Napoleon at Waterloo* has an elusive solution. The basic tactic for the British is to attack. But after that, the game has a number of solutions.

One of the advantages that board games have over miniatures games is that they are potentially much simpler. This simplicity is one of the board games' chief virtues. Until there is a strong enough market to support more complex games, a designer must accept this simplification. To this extent, this market seems to be developing. The state of the art has been advanced to the point where rules previously considered complex are now considered *de rigueur.* Thus the designer must consider this new direction in games as well as the myriad other decisions he must make. The options are many; it is up to the designer to choose.

RESEARCH

Research is the key to designing a good

simulation. A game *can* be done without research, but any game which has a pretense of being a simulation of an historical event has to contain the historical element. This means the designer must be fully versed in the history of the situation he is simulating. A designer must do a lot of spadework, keeping on file a lot of essentially trivial information for which there might be little use in the immediate future—or perhaps no use at all. What appears to be useless information on a given period of history may well save some steps in research for a future game.

For many people, the favorite question is, "Where does the designer find all of his information?" A designer really cannot do historical games unless he is either an historian or has an extremely well developed interest in the field. Even a well-written book about a period will not readily furnish all of the information necessary to do a game, and the older the era the truer this becomes. This is why experience is so important in running down the more obscure points not touched upon in general sources.

A designer doing his research has to recognize that once he gets into the pre-World War I era he is beginning to deal with different styles of writing history. Things that modern historians take for granted when they write a detailed history of a period are simply ignored in older histories. Added to this is the fact that military history is a relatively new field. Prior to the American Civil War, no war was really covered extensively. Individuals might write their memoirs, but detailed histories of wars, from anything like an objective standard, become fewer and fewer the further back one goes. A curious point to keep in mind is that the reverse is not always true: the more recent the event does not mean that the history will be as thorough. Almost all of the participants in World War I have written official histories. Most of the participants in World War II have yet to do so. The U.S. Army is still turning out volumes on the war. The book on the Rhineland Campaigns was only available in 1974, 30 years after the event and they have not finished the whole War in Europe phase, not to mention the technical services. Of course, no official German history has been written, nor is it likely that any will be because so many records were destroyed in the final days of the war and so many more were carted off by the victors. It is doubtful that anyone can even know what records actually exist.

Another factor to consider is that researchers are regularly turning up new facts on past wars, wars which one would normally think were exhausted in terms of information. Just recently revelations in Britain about the Ultra device, which read the German codes, make obsolete many of the speculations about Allied actions. There is even new evidence emerging on such "far-away" wars as the American Civil War! Events such as the Bicentennial will produce a more thorough—and less romantic—view of some of the older wars, as historians, anxious to take advantage of the topical nature of the event, devote more attention to it than before. A rash of good books recently emerged (1966) dealing with the Norman invasion of England, the 900th anniversary had provided a good occasion to review the invasion extensively. This sort of new information about old wars adds another dimension to the researcher's problems—he cannot assume that simply because the writing is contemporaneous with the event that it is the most accurate or thorough account. Often it is, but sometimes something much more recent is better. Therefore the designer must have a good grasp of the sources available.

Research must be thorough, and that means that it is time-consuming. Unless it is a subject previously researched all of the work must be done from scratch. A typical SPI game will often involve 100 man-hours of research, or $400 to $500 worth of time. For a company which is trying to make a business out of publishing games, this money is a factor which must be considered in determining total costs.

When a designer is doing several games on

one subject, whether the level be strategic, operational or tactical, he can sometimes save money by transposing the information available, as happened, for example, with *Desert War* and *PanzerArmee Afrika.* There have been many books written on tank warfare in North Africa and so research was fairly easy; spread over two games it was also inexpensive. Research for SPI's divisional level East Front series *(Moscow Campaign, Panzergruppe Guderian, Turning Point, Kursk* and *Destruction of Army Group Center)* was aided by the fact that the Germans actually published their Order of Battle (OB) for the entire war. This obviously takes some of the time out of research and permits the time to be spent on other phases of the game. Thus research done on one game aids another.

When dealing with an earlier period the designer has an advantage in some ways. The necessary information will simply not be there (or, if it is there, it is in something so obscure that neither the designer nor 99% of the people who play the game will ever have heard of it). This permits the designer to take a "best guess," at a given situation. Many designers have been fairly successful with "best guessing." For example, nobody really knows how the Greeks and Persians actually fought. There are a few contemporary reports, but they are skimpy and sometimes not at all reliable. Contemporary efforts to reconstruct the Greek and Persian techniques often differ. There is still debate, for example, on exactly how the Romans fought in formation —did they leave open spaces or did they pack their men tightly together? Nobody bothered to keep notes on individual swordsmanship because a lot of the training was simply oral tradition. Recent games such as *Alexander the Great* and *The Conquerors* used a great deal of "hunch" research in reaching their "game" conclusions. The designer has to be prepared to deal with these "not sure" and "guesstimates" and "best guess" types of information. If he has any inhibitions about doing this he is going to run up against one wall after another because the hard facts are simply not there to answer the questions in a definitive fashion. In the end, he won't get the job done.

The researcher has to recognize this. He has to recognize that the official system is often not what was actually used. Viet Nam provides a modern example. The Army had a very fixed organization for an infantry battalion in terms of men and equipment. It is safe to say that almost no unit, after the first few years, ever fought with exactly the authorized amount of equipment—no additions and no subtractions. This makes some of the activities hard to understand for someone who knows only that the Army has a fixed organization and equipment authorization.

The sources for information can be quite varied. (See the Chapter on Research Sources). For modern wars, a good place to start research is the *Reader's Guide to Periodical Literature.* Then there is always the problem with propaganda. The older the period, the easier it is to recognize propaganda. Propaganda does not age very well; exaggerated claims of friendly success and enemy failure tend to become apparent when the general course of the war is known. When reading contemporary American accounts of the Battle of the Bulge it is clear that they are trying to minimize the impact. They admit the attack is taking place and, generally, they reflect an accurate picture of the situation as it developed. However, they do not talk about the initial disintegration in the American positions nor the fact that the Americans were caught off guard. This type of propaganda is readily recognizable now.

An interesting rule of thumb to keep in mind about current games is that our knowledge concerning current military events is not that much inferior to our knowledge of what happened twenty years ago. Not much more information comes out of any current event than is available six months after it occurs. For example, during World War II a number of books were put out that contain about everything found in books published since then. Some of the finer details are missing, but, aside from something like the Ultra/Enigma revelations (which were unique in nature), nothing substantially new has surfaced.

For example, SPI was able to obtain a great deal of information for its *Sinai* game from *Time* magazine's coverage of the 1973 Arab-Israeli War. A lot of game research had been done on the 1967 war and a hypothetical 1970's scenario. *Time* and *Newsweek* filled in the gaps and, judging from comments received by SPI from Israeli officers, did as well as the histories which have been published since then. The game pretty well hit the nail on the head using immediate sources.

Research is not a spectator sport. The more experience a designer has, the more rapidly he can do the research, and the more accurate it will be. When doing games on a given period, there is a lot of background—economic history, social history and politics—which must be absorbed, even though it may have no direct effect on the game. All of this is done to get a feeling for the period. For example when dealing with the Romans, people tend to think of them in modern terms, but the key to a good ancients game is to start thinking in Roman terms. When dealing with the past, a designer is dealing with a different perception of the cosmos and environment from the one found today. In the feudal period, the designer has to consider why anyone would want to go on a Crusade. What does it really matter who controls Jerusalem. It didn't really matter to the Christians for some four hundred years before the Crusades and it hasn't mattered since then. Why, during that period, did it suddenly matter? In such a game, there can be no doubt that one requirement for the Crusaders is that they hold Jerusalem. This becomes a victory condition.

Basically the researcher is looking for motives which may not be comprehensible to modern people. People deal with a specific set of values at any given time, but those values change throughout history. For example, at the end of the American Revolution, a lot of people wanted to make George Washington king; no one wanted to do the same for Lincoln seventy years later. The concept of states' rights was a lot different in 1789 than it was in 1859, and the present view differs from both of them.

A game designer must be familiar with, and thus design the rules for the period. If he tries to do a Marlborough game using Napoleonic concepts, he won't be doing a Marlborough game; he'll be doing a Napoleonic game. Marlborough cannot march down the Rhine by living off the land in Napoleonic fashion. If he wants to march down the Rhine, he must perform what was, in the early Eighteenth Century, a logistical miracle.

Another problem to consider is non-combat deaths and infirmities. The majority of casualties prior to World War I, were *not* combat casualties. This type of attrition has to be considered. It was a typical experience in the Napoleonic period for a leader to start in the spring with a corps of about 60,000 men, fight no major battles; yet end the campaign six or eight months later with 14,000 or 10,000. It was normal and expected. American history makes a great deal about Washington's problems with deserters reducing the ranks of his armies. The British had the same problem. It was reduced only because the Americans were at home while the British were in a strange and hostile land. A designer must make himself aware of this type of problem, in order to be able to use it—if the situation calls for it—in his game.

Sometimes research will turn up some surprises, or at least information not widely known. For instance, the North Vietnamese had a problem with non-combat related attrition. They had a much higher malaria rate than the Americans. Research revealed that the average North Vietnamese soldier was not at all used to living in a jungle. They did not have jungle sense the way everyone assumes the Japanese did in World War II, though people have come to view the North Vietnamese as quite at home in the jungles of South Vietnam. Actually, the Japanese had no special affinity for the jungle either.

They were city or farm boys and actually fared worse in the jungle than the Americans. The Japanese lost about five percent of their army a week to disease on Guadalcanal. The Americans might lose half a battalion to disease in six months; the Japanese would suffer ninety percent casualties.

Such important, albeit not well known, facts are the elements which make the difference between a game and a simulation. They are one reason why a designer needs a wide range of knowledge about a period. At the same time, a person designing games requires a special approach to military history, what might be called a "total approach." The designer must consider the human factors—whether people got sick or didn't and what impact this had on the course of events. The rules and concepts that make up a really realistic game can be grasped only if the designer has a total understanding of the period.

A game designer cannot separate military history from history as a whole. When he does that something is lost, and a lot of people who play the games realize this. Granted, games are pretty much military in appearance, but there are a lot of things built into them which are not military. All factors have to be considered. Of course, if a designer has developed a generally broad knowledge of a particular aspect, it makes it much easier to go to another period.

The key factor in taking advantage of all of this history is the analog. The human mind is, in effect, an analog computer. An analog computer is one which compares. In studying historical things, it is also possible to use the analog. Because minds are complex, very elaborate analogs can be made. For example, in the Balkans of the Eighteenth Century, there was a great deal of partisan warfare. There were the Croats and Pandors and all types of nasties running around. Immediately the designer should make an analog: "What similarities does this period have with ones I know more about?" By using these analogies, he can say: "This doesn't look just right. I'd better check this particular aspect. Just what was the degree of religious intolerance in this period?" Once this is accomplished the designer has made an adjustment between the Balkan situation and, say, the Ukraine in the Second World War.

Regarding specific methodology in research, it is a good idea to start by using general references to get an overview of the situation. However, after a while the designer has to start going to more detailed sources. Quite often he'll get caught on one point or another only to resolve it by catching one little item, such as statistics on disease or attrition, or the fact that the militias of the American Civil War were actually a major portion of the manpower available to both sides during the war.

One of the big problems a designer has in doing research is that he often doesn't know just what he is looking for. SPI's *Tank!* is a perfect example of that. Tank weapons require far greater ranges than the weapons used in *Sniper!*—out to three thousand meters. If the game were to use a ten meter hex, that would mean that 300 hexes would be needed to show the entire range of a tank weapon. For some people, it takes a great deal of research just to learn that point. Obviously, other people can take that information out of the "data bank" they have assembled from past research and past experience, avoiding the time needed to determine this crucial bit of information. However, if the designer keeps on digging, he will find that most of the World War II tank combats took place at less than 500 meters' range. That means the most of the area needed to show the maximum range of the tank can be ignored.

Another thing a person doing an individual tank level game has to consider is the type of formations used. Oddly enough, one way to do this kind of research is to look at photographs. Although field manuals will show the official way of doing things, practice usually differs from theory and units will generally fight in the same formations they use for maneuvers. There are a number of photographs of units on maneuvers—a great deal more than show units in actual combat formations. These photographs serve to give that piece of information to the alert designer. On the other hand, the designer can run into some problems because there aren't too many accounts of what tank against tank combat was actually like. The designer will have to use the "general" situation and be compelled to cut a slice out of that.

A large organization such as SPI, which does a number of games, can cut short much of their research time because of trade offs from other games. As mentioned before, the information gained in researching one game should never be discarded as it can be useful, later on, for other games.

The types of sources to use in research will often vary, depending on the type of game being done. While the official histories will be invaluable for strategic level games and many operational level games, the tactical games require different sources since official histories rarely deal with small unit actions, even in critical battles. Aside from field manuals, which present the problem of doctrine not agreeing with practice, the only other source readily available to most designers is popular histories and personal memoirs. The problem with books like that is that they are not written with much rigor. They tend to glorify the writer and emphasize certain aspects out of proportion to their real impact. On the other hand, the official histories usually compensate for the lack of rigor in adhering to balance but lack the "flavor" which is often what a designer is trying to capture in a game. Once again, experience is important. When dealing with these personal histories, the designer has to fall back on experience and analogs. A general principle in researching a principle game is to get the best information available and then make an educated guess. Actual sources for use are covered in a later chapter on Sources for Research.

Making an educated guess may not seem very professional, but that depends on how it is handled. An educated guess, in the hands of someone skilled at using the technique, becomes more a case of making valid assumptions, rather then a shot in the dark. A game designer needs to know how strong a unit is in assigning it a value in the game. An historian may want the same information but if he can't find it, he can simply gloss over the fact. He can do it in such a way that the loss will not be apparent to the reader. But a designer cannot simply omit a unit because he doesn't know its strength. He must make an educated guess as to how strong it was. In this respect, the game itself can be of some help. The historical result is known. If the game produces the same historical result, the chances are reasonably good that the guess was correct. The exact strength of Hannibal's forces at Cannae is not known—nor is the exact Roman strength when all of the allies are counted up. But if the Carthaginians can defeat the Romans employing the double envelopment tactic, then whatever reasonable assumption is made is probably a valid one.

By contrast, in an area such as the American Civil War there is a great deal of information and often it is simply a matter of finding the right books. However, in a war like that the sources are sometime too abundant and often contradictory. It becomes a problem of separating what is needed from what is available. This can be difficult because the designer sees so many possibilities and is tempted to include them all. He can design three or four different games on the same subject, each emphasizing a different point. With this type of mass information, the designer must set limits for himself and enforce them. Otherwise it is easy to get sidetracked.

Similarly, where there is a lot of contradictory information on a given event, it is difficult to decide which version to use. E.g., the versions may have charts and diagrams which appear to be authentic but don't agree. The designer may be lucky enough to find a source which is more impartial or, the sources might admit they are estimates but the estimates vary so much that no median is readily apparent.

It is important, when researching a game, to have goals. The game designer should use the technique of researching by objectives. A researcher should always be on the lookout for information in unlikely places. *Scientific American,* for example, had a detailed discussion of ancient slingers and their effectiveness. This information might not seem to be readily usable, but it also is not found in any of the likely sources to which a designer would go in researching a tactical game on ancient warfare.

Flying Circus originally started with trying to resolve the problem of depicting a three dimensional situation. It was solved through a very thorough analysis of what went on in air combat. This is usually the easiest way to a solution for any game problem—how did it actually happen. A designer needs all the details, as he never knows when first doing his research what detail is going to be the critical one.

There were a number of critical factors in World War I air combat, all of which impacted on how the game came out. First, there was the fact that air to air combat was decided primarily (90%) by one plane getting the jump on the other. Another problem in World War I combat was that the planes were extremely primitive. They suffered as many operational losses in taking off and landing as they did in the air. That became an important factor in dealing with the ability of planes to dive in the game. The limit on diving is not how fast the plane itself could go but how fast it could go with its wings still on. One of the reasons why the Fokker DIII triplane was only in action for about four months was that the pilots found a critical flaw which they could not correct. When the pilot went too far into a dive, the tail would come off. It took an extremely expert pilot to know just how far he could push it. (Richthofen was one of the few pilots who could handle it.) Most of the pilots were only too happy to give up their DIII's for the new Pfalz or Fokker DVII.

The question of altitude, in *Flying Circus,* was not critical when maneuvering. Once a pilot was on the same level as his opponent, combat was a matter of a dog fight, primarily involving turning. Most flights evolved from the very simple idea of either out-flying the enemy or just out-turning him. Because of this, the designer determined that the altitude levels could be made very small in number. It wasn't necessary to have one for every ten feet. One for every 500 feet was satisfactory because these planes climbed fairly slowly. They could climb at a rate of perhaps 150 feet in a minute. For all practical purposes an enemy plane 400 feet above his enemy was immune—the plane below could not get to him. Nor could he shoot at the plane below because of the short range of the weapons. To climb to the higher level would require sacrificing all lateral and horizontal movement while the enemy just moved away. Recognizing the time and space parameters of World War I air combat gave the key to solving the altitude problem. Actually, the altitude problem was solved back in 1968; the final development, using markers, was a logical extension.

A lot of data has been published on unit losses, but none of it is directly applicable to games. The most common reference is FM 101-10, published by the Army. This is the Staff Officers Field Manual. It is mainly based upon World War II and Korean War experience, with an addendum on Vietnam. It gives the attrition rates for divisions in terms of percentages of manpower lost per day for various operations: attack of a fortified position, attack, period, holding action, pursuit, retreat and delaying action. The problem this information presents is complex. A division might show a loss in a given operation of three percent but most of these losses are by the infantry battalions. As much as ninety percent of a division's casualties may come from an infantry battalion. With a division of about 16,000 men, three percent casualties mean 500 men lost a day. But most of that 500 is coming from the infantry. How many battalions will be engaged in this type of action? How many battalions are in the front line? This information is not readily available. It means reading other sources to find out the answer to that question. It might be that the majority of the infantry losses are being taken in only five battalions. That means a hundred men or twelve percent of a battalion's strength is being lost a day. Of course, a battalion does not decline on a straight line basis in terms of its total manpower. Most of its losses come from the combat elements, not headquarters and logistical elements. A loss of twelve percent might easily represent twenty-five percent of a battalion's effective combat strength. So when the manual says three percent of a division it may mean about a division quarter of the combat strength of five or six battalions is being lost a day. Obviously, it could also mean those losses were all taken in rear echelons due to artillery fire. The manual is merely dealing in averages and, as a result, is a deceptive source to rely upon too heavily. It *is* a good source for "guesstimation" information, as mentioned earlier.

A designer can make a good game without doing his research. The difference between a game and a simulation is the degree to which it reproduces history. No designer can approach a simulation without recognizing that he has to do his homework; he has to spend the time to become thoroughly versed with the period, not only from a strictly military point, but also to learn the atmosphere in which the given fighting took place. With that, he can take the first step toward a good simulation.

QUANTIFICATION

Quantification, as used in the game design context, consists of taking known elements, such as unit size, mobility and fighting ability, and reducing these elements to the numerical values usually found on playing pieces: the combat strengths and movement allowances. The same term can apply to any element of the game in which raw data are reduced to game terms. Quantification is the heart of game design. It forces the designer to make a decision; he has to say that a given unit is worth this much in relation to another unit. And most times quantification is just an educated guess, and that's the whole secret.

There are two ways to quantify the basic elements. The true guessing aspect is used either by the very naive designer or the very experienced designer. In between there are the numbers. In other words, there is the analog approach, which is a fancy name for guessing, and the digital approach, which involves counting rifles. All designers have to go through the counting phase, because the experienced designer's "guessing" isn't just a shot in the dark. It's based on a lot of experience, but he will also spot check to be sure his guesses are correct. This guessing system does work, and of course it saves a lot of time. It took about three or four hundred hours with *Jutland,* and perhaps a hundred or so with *1914,* to do all of the "counting." (This includes finding out how many guns of various calibres the fortresses had, and what the rate of fire and ammunition supply of the German siege mortars was.) A professional game designer can't afford the luxury of spending a hundred hours counting. However, an amateur could, because it's his own time; he doesn't depend on the game to put bread on the table.

There are a lot of little tricks involved in the guessing game. Many people use them unconsciously. For example, there is the practice of using another game as a reference point. Any game based on some system, such as the games based on the *Kursk* system, basically uses this concept. Ultimately someone had to sit down and figure out what each unit represented in exact terms. Once that was done, it was easy to expand. *Red Star/White Star* was done laboriously but it furnished the basis for *NATO* and *NATO* was done by analog. Of course there were some variances due to the scale of the game but they were not too far out of line. From *NATO* came *The East is Red.* The Russians were quantified from *NATO* and all that remained was to analogize the Chinese to the Soviets, based upon the research which had been done.

Terrain is another element which the designer must quantify. Quantifying terrain is a help rather than a hindrance. A designer can cover up a lot of things in terrain as nobody every really looks at terrain. They see the Terrain Effects Chart and they follow it religiously. Nobody really argues about it that much. Nobody gets worked up about swamps costing four Movement Points instead of three or something like that.

The Terrain Effects Chart is often a useful place to camouflage certain quantification defects. It must have a modicum of realism, of course, since it is obvious that some terrain slows an army down more than others. Some terrain is impassable while other is barely passable. On the other hand, terrain can prove a major stumbling block if a designer is not wary. For example, in *Destruction of Army Group Center,* there was the presence of the Pripet Marshes, which the Soviets managed to cross relatively easily. They had what they called mechanized cavalry groups, a combination of cavalry and some mechanized infantry. The problem became how to build something like this into the game. The designer didn't want to create a whole new series of special units and rules. He could have created units with special movement abilities only within swamp hexes and something else outside of them, but that would have become very artificial and awkward. The solution proved to be not at all sophisticated, if a bit of a cop-out. Movement through the Pripet was made very hard for all units, and these special units were basically ignored. It was really a matter of designer's bluff, and it apparently worked because nobody commented on the inability of the Soviets to do what they did historically. In part the designer was helped by the fact that the campaign is not well known, though it should be better known for its impact since the Soviets literally destroyed an army group, and the average player didn't know that the Soviets did this type of maneuver through the marshes. In the play of the game it was a minor point and it wouldn't have added much to the play to have three or four units able to go more readily through the swamps.

Any regimental level game on the Ardennes raises a fundamental problem of the geography of the region. On the strategic level the road network is often taken out. This is a situation where the problem comes out in the wash because the scale is so large as to make the road network of little importance. But on the regimental level, the roads become significant. Avalon Hill's *Battle of the Bulge* game was one of the first to have a complicated road system. This game also had some tricky, critical terrain. Their solution was to ignore certain critical aspects of the terrain, simplify others and leave others out. The other extreme was SPI's *Bastogne* game, which included everything. Neither approach is very good though the Avalon Hill version, considering all of the inevitable problems inherent in that scale of game, is probably better. The best, though still faulty, solution is SPI's *Ardennes Offensive,* which simply ignores the secondary roads, goes a little easy on the movement restrictions off roads, but shows all the main hard surface roads.

Terrain effects should not be a significant *design* element. Terrain is either passable or difficult to move in, but that's all that need to be worried about. Most terrain is neutral in that it affects both sides equally. While terrain should not be ignored entirely, neither should it be a critical design element which can be used to swing the game one way or another. The SPI preference is to keep terrain neutral in its impact. Whether it should be used that way by everyone is really a matter of personal style.

The next element to be quantified is perhaps one of the hardest: combat resolution. Combat resolution requires comparison of the two major quantifications, terrain and basic combat strength. The Combat Results Table (CRT) is an estimation of the chances that a given result will occur, at least in its initial form. An effective system is to use the old analog technique. Make a basis, for example at three to one or two to one and then determine that a certain side will lose so many points for every six outcomes. Then assign weight to effects like "defender retreat one" or "two" or "exchange" or "attacker retreat" and so on. Everything is based upon an analog. Once this analog is made, extrapolate it to make the full CRT. This is fairly simple to set out and, because it is simple, a designer should have no qualms about going through three or four or half a dozen CRTs in a single game until he has the one he feels is right. A CRT will, in effect, design itself. It will either be right or it will be wrong. Put it to the test and a designer can find out very quickly whether or not it is working. There is really no off-the-shelf way of determining what the attrition rate should be for a CRT. It looks so easy, initially, but the problems become apparent the longer the designer works with it. Unquestionably the CRT poses the most problems in the area of game design.

Another problem with games is that the designer is trying to create, in written form, a mathematical relationship—an algorithm—which is what the rules and Victory Conditions are about. If one of these mathematical rules has a long formula and just one thing is changed—a plus to a minus—this alters the whole result of the formula. Games are very susceptible to this. It's a testament to the abilities of most game players that they aren't overwhelmed by this problem. They usually know when to change the minus back to plus and when to multiply instead of divide if the designer has made an error in his formula. However, there are limits to what a novice can do in this area. This is most obvious when a novice proposes a rules variation to correct an imbalance which he perceives. All too often that variation has the pebble-in-the-pond effect: the first ripple may be what he wants to change, but he forgets that the ripples keep on spreading and affecting other elements in the game. Often, to bring one section into line, the novice will further unbalance the rest of the game.

When a designer decides to turn a situation into a game, he has to run through all of these elements. Some people do it better with a check list. Some prefer to keep "running the tape," so to speak, in their heads and eventually everything gets covered. When the designer decides whether a game will be simple or complex, the designer then takes into account the availability of information. He can do a lot of these steps at once. He can look at a game situation, size up the scale that should be used, the amount of information that is available, the amount of research to be done and render all these decisions virtually at once. But no matter how it is done, the designer will end up doing some research, even if it means using only a few reference books. This research can also be complex. For *CA,* SPI ended up doing a lot of mathematics to figure out the acceleration and deceleration factors in steam-driven ships.

Victory conditions are another element of quantification. They are often very difficult because it is frequently unclear just what the victory was. Obviously, Napoleon lost at Friedland and won at Elyau. But there is winning and then there is winning. Take the situation involved in *Lee Moves North.* Historically, the best that Lee could hope for was a political victory, and to do that he had to pull troops out of the west. The most he could hope for was to frighten the North. He couldn't do anything that was really damaging. Therefore, victory at Gettysburg might have achieved the goal of frightening the North but it would have had little impact in the sense of "leaving the North open to conquest." Therefore a victory for the South in *Lee Moves North* must consider that the best he could hope for was a local victory, not conquest of the Northeast.

One of the obvious abstractions in games is in movement. A player can move one unit its full movement allowance, stop in a zone of control and attack an enemy unit while another friendly unit moves only one hex and attacks an enemy unit. Theoretically, the unit which moves the shorter distance is taking less time. Most games ignore this differential, though some require a movement point cost to enter a zone of control. The point is that a movement allowance is a quantification of time and space—the amount of space a unit can cover in a given amount of time. What actions a unit can do in that time/space abstraction depends on the time each game turn represents and the distance involved in each hex, coupled with the designer's personal view of how much should be accomplished during that turn. This, again, often requires additional research into fighting techniques of the time involved. In reality, the time of combat in a fixed period of a day or a week is usually rather brief in proportion to the time spent moving and waiting for combat. At the tactical level, where a game may have hour long turns or less, even then there are relatively short periods of fighting.

Morale is a difficult factor to put into wargames. The miniatures people have death with morale for a long time and it has always been one of their big gripes that boardgames as simulations, traditionally, did not take morale into account. However, recent games such as *Wellington's Victory* and *La Bataille de la Moskowa* have gone a long way to remedy this. Quantifying morale is simply making a numerical determination at what level a unit will cease to be effective. This information is often subjective, however, and at best is difficult to find.

Quantification, then, is a direct relative of research, the two go hand-in-hand, together with a good grasp of game design principles, or the integration of all available information.

INTEGRATION

What people normally think of as game design would better be termed *Integration.* This is where the designer takes all his research and ideas and actually *turns* it into a game. *The American Civil War* (SPI) serves as a convenient example to use by way of illustration, particularly to show some of the options available.

The introduction to the rules says: "*American Civil War* is a simulation of the American Civil War between June '61 and June '65. Union and Confederate units represent the actual forces which took part in the fighting." This type of introduction is in every set of SPI rules but it actually is more than a simple introduction. It defines what the game is going to be all about. You have to decide that before you can even start, and that is the first thing the designer should consider. The crucial phrases here are "between June '61 and June '65" and "the actual forces." The game represents the whole war, outside of commerce raiding on the high seas.

Turning to rule 2.0, General Course of Play, "Civil War is basically a two player game. It is played in sequence turns." Those two sentences convey two important elements of the game. The game could have been a multi-player game; it could have been a simultaneous movement game; it could have been a solitaire game. Those sentences reflect the decisions made by the designer and, as can be imagined, they have the widest repercussions on how the game turns out. They are so obvious that many people would not consider them to be the critical decisions, which they actually are.

The rule goes on " . . . during which players move units on the map, have combat, achieve certain objectives. Victory is based upon the Union achieving certain objectives by a certain turn in the game." These are all decisions that have to be made by the designer. Each of these initial paragraphs defines another aspect of the game. The designer, in setting them down, is laying down building blocks for the actual game.

Continuing, "Each game turn represents three months of real time. The entire game lasts at most fifteen game turns. . . . 2.2. Game scale. Each hex represents 25 miles of land from side to side. Each strength point represents 5,000 to 10,000 men, depending on the quality of the troops." These statements convey an outline of the structure of the game.

These key elements were decided upon because of various factors. The elements of the introduction, 1.0, were obtained from the feedback question which prompted the game. Several decisions had to be made in formulating the feedback question since, if the question were rated high enough to warrant doing the game, SPI would be bound to deliver a game which at least met the specifications of the feedback. A two player game was decided upon for several reasons. First, a two player game is the easiest type to do. Second, *The American Civil War* employed a somewhat new system, a fact which was expected when the *feedback* question was formulated. The system was only loosely based on the *World War II* system. The Second World War was on a much larger and more sophisticated scale than the American Civil War, and this made for critical differences. As a result, the *World War II* system could only be very loosely adapted. The game could have been a solitaire game and could have been an excellent multiplayer game.

A designer plays a mental game with each sentence in his rules. He makes a simulation in his head as to what factors bear upon each element. This is all done very deliberately. There is very little luck involved in designing games, although there is serendipity. But a designer can't depend on serendipity to save the situation often enough for it to be considered a legitimate tool.

The biggest constraint against *American Civil War* being say, a simultaneous movement game, was the size of the conflict: it covered an entire war and a large amount of territory. The designer thus has to be aware of the limitations certain systems carry. Some games are too complex to be two-player games; others are too simple to warrant multiplayer situations, even though several different forces might be represented on the board. And systems like simultaneous movement can only apply in restricted (usually tactical) situations.

The rules go on to say: "Victory is based on the Union achieving certain objectives by a certain turn of the game." This reflects the historical and game-fact that the Union had the initiative. The South will win by default if the Union does nothing. For example, Lincoln had to achieve major progress by 1864, when he was up for reelection, or he stood a good chance of being voted out. His opponent in 1864 was McClellan, one of the generals he had fired for inability to do the job. The importance of military success on the political campaign is reflected in the game.

An essential element of integration is to create as many moveable "parts" as possible, so that it is easier to adjust an element or fine tune the game without doing a complete redesign. Unfortunately a lot of people put a game together in a haphazard fashion and leave themselves no room to move. They force themselves into a very stiff game system which doesn't allow for any slight adjustment. And quite often all a game needs is a slight adjustment to make it work well.

A lot of decisions on the integration of a game are reached after days and weeks of considering the problem. A designer should not be too hasty in making up his mind. On the other hand, a professional designer has to remember that the clock is always running on his design and the longer it takes to get it out, the more money it will cost. Ultimately, whether professional or amateur, the designer has to come to the point where he will make a decision. While putting in too many bad decisions can reach a point of diminishing returns, at the same time the wargame designer has to take some chances with his decisions if he is not sure. The point is that the designer has to juggle the two requirements of precision and meeting anything like a time-table. One or the other will have to fall from time to time and the designer has to decide which it will be.

Realism is an important goal in integration. The designer is not making merely a game; he is also rendering a simulation of history. On the other hand, realism is not necessarily the same as historical authenticity. If the player can do what actually happened, that is a realistic game. If the actual events were not the soundest and the player deviates from them or if the game clearly indicates that the better course of action for a player is to ignore the historical results, that does not make the game any less realistic. The game *World War II* has been criticized as being unrealistic. This simply isn't true. The Germans can do (or attempt) everything they did in reality; they can fail to do everything they failed to do (such as leaving Gibraltar alone). On the other hand, the game makes it pretty clear that if they waste their strength on these ventures, they will probably lose—as they did in reality. If the German player wants to win, he has to adopt a different strategy. That will vary from player to player and may produce some strange results, such as occupation of Turkey. That makes it no less realistic as this was an option open to Hitler which he never exercised but which the Turks were certainly concerned about. On the other hand, certain factors were eliminated for simplicity, such as the Allies invading neutrals. This would have involved adding a whole range of political considerations and politics were played down in the game.

One reason why integrating a game excites so many people is that it is the last stronghold of the individual. All the other aspects of the game, even research, and rules writing, can be done by a team. Unfortunately while integration may be the easiest thing to do in game design, it is the hardest to do well.

The ability to analyze a situation is a principal key to integration of a game. The designer has to dig into his research material to determine what should be factored into the game and what should be excluded. This is where experience comes in, as it requires extensive analysis to smoothly integrate a game. During the integration phase there is a sort of game interaction going on. The research information is actually producing other (game) information. If the information produced seems to be wrong, then the integration may be wrong. On the other hand, sometimes this information production can actually fill in gaps that the research could not cover. Again, recognizing when false information, indicative of erroneous integration, is being produced—as opposed to good, valid and new information—is an experience factor.

In deciding on whether to use simultaneous versus sequential movement, it is important to remember that simultaneity becomes more important the lower the level (i.e., tactical v. strategic) of the game. At a tactical level the player is dealing with a situation where an order is to be executed immediately, not 60 minutes or three days from now. As far back as *Tactical Game 3,* the forerunner of *PanzerBlitz,* it was apparent that tactical games were popular. Yet, SPI could not produce a game like *Sniper!* using a sequential movement system. The sequential system does not convey the sense of immediacy which everyone knows occurs in a small unit engagement. This meant making a quantum jump and solving the problems of simultaneous movement and all that it brings along. This is the area where an amateur is at a big disadvantage. Merely playing games frequently does not bring the knowledge necessary to advance the state of the art. Advancing the state of the art is sometimes the only way to resolve a particular problem in a satisfactory fashion.

The designer has to use his discretion when integrating a game. Someone who knows too much about a subject is tempted to put everything he knows into the game. One of the reasons *Fall of Rome* became a problem was that the designer wanted to put too much into the game. When a chance came to do "Son of Fall of Rome" for *Horizon* magazine, (they paid for it, but never published it), it was radically reduced in complexity by weeding out all of the extras. A designer who knows all of the trivia has to resist the temptation to show it all.

The Army helped, indirectly, with the *Red Star/White Star* game by publishing their field manual FM 105-5, *Maneuver Control.* The Army assigns a fire power factor, in the form of a numerical value, to every unit from the individual weapon on up. They just multiply to get the higher units. They consider all of the weapons that are assigned to the various units for both the Americans and the Russians. They even include range attenuation as somewhat of a simplification, showing that even the Army is forced to simplify their quantification. The application of these firepower factors is different in the Army system from most board games. Casualties are a percentage of how far and how long a force is advancing through various fire power zones. The concept, even if a designer chooses not to follow it, is a good place to start.

The biggest problem in doing something different is that there is so much latitude for mistake. When they start designing games, a lot of people realize very quickly that it's easy to mess up a game. A fledgling designer can get discouraged by this. He spends all this time on his game, and then it doesn't work. The problem is that he has forgotten how he put it together so he doesn't know how to take it apart for "repairs." This is another element of designing. A designer must be able to recall just how he arrived at a particular point.

In integrating a game, a designer must decide which direction he is going to take. Once that is decided, this begins to lock him into a lot of decisions. He has relinquished control over a number of elements. The further one goes in designing a game, through integration into development, the fewer decisions can be made. A decision in the first week has more impact than ten made ten weeks later because the first decisions are so significant in their scope. By the time he is at the end of the game, he is reduced to such decisions as what colors will be on the map. The decisions become less and less significant the further one goes into a game.

Fortunately, games have an element in them that allows the designer to validate whether the game is true; i.e., whether the game produces the right results and meets the interim "bench marks" along the way. This is a factor that limits the designer as to how much he can tinker with his invention. Some games in the past have violated this precept. For example, *Stalingrad* and *Afrika Korps* had correct final objectives in that the end result reflected history. But the intermediate objectives were almost completely lacking in historicity, and the games had little value as history. Thus, it is with the integration of his information that the designer shapes the destiny of his product; his decisions will reflect his information and his ability as well as his biases and prejudices. His ability to control all of these elements is the measure of his integration. And integration is the measure of the game.

RULES

It goes without saying that the rules are the backbone of any game. However, it is hard to comprehend how difficult it is to write a sophisticated, understandable and playable set of wargame rules until one actually tries it. While there are myriad people who could design a game with minimal trouble, there are probably less than a handful of designers who can produce acceptable rules today. And, perhaps even more curious, the better the rules the less they are appreciated.

What exactly goes into rules writing? Well, the first thing that must obviously be considered is the content: what is the game about? While this may seem almost inanely obvious, it is important to consider what you are writing about. The professional designers have learned that different subjects demand different approaches. For instance, the tactical air modules for operational-level games (such as *Drang Nach Osten, Coral Sea, Bar Lev, Jacksonville,* etc.) all seem to follow the same pattern, even though they were designed by different people! Thus, if a prospective designer is familiar with other games on his topic, then he will find it easier to get into the style and language of that topic.

And style is another important function of rules-writing. Do you want to be conversational and smooth, like designer John Hill for his various games, or do you want to be definitive and legalistic, as in the approach taken by SPI. In the latter instance, SPI has, through a great deal of trial and error, adopted a Case System whereby each rule and each subsection of that rule is numbered. Thus, the basic Movement Rule might be a 5.0, while Restrictions on Movement would be 5.1, and a discussion of the effect of mountains on movement, say, 5.13. This, combined with a penchant for using precise terminology and *repeating* that terminology from one game to the next, produces a highly refined and legalistic approach to rules writing.

Now, although SPI feels that its legalistic approach is best—and it certainly has many major advantages over its competitors' styles—there are many gamers who dislike this form of writing. Legalistic rules tend to be rather dry, and they are certainly harder to read than a folksy, conversational set. Moreover, poorly written legalistic rules can be virtually impossible to fathom. Yet, there can be no denying that the amount of information contained in the rules, the facility in finding specific rules and cross-referencing them with other rules, and the simple fact that the rules leave little to the players' imaginations all are major advantages to consider.

Thus, the potential designer/writer must decide what approach he will use: what audience is he aiming at, and, more important, what style will he be comfortable with? Most fledgling designers take the conversational approach, as developing a legalistic system takes a great deal of time and effort. Of course, there is no harm in copying the general method used by a specific set of rules you admire; using a professionally designed game as a model is not only *not* "underhanded", but it actually makes good sense! In any case, the final decision on rules format usually rests with the developer, as he is the person who writes the rules out.

There is often a great deal of controversy over whether a specific set of rules should be simpler, etc. These questions usually apply to a game that is, by its nature, at least moderately complex. SPI has received many complaints about its naval system in *World War III* and *Global War.* A majority of those complaints centered around the fact that these rules should have been written more simply. The fact of life is—and it is a lesson that should be taken to heart by the amateur designer right from the start—that complex ideas produce complex rules. There is no way of phrasing a multifaceted concept in two lines. To do so is to invite trouble.

And the trouble you are inviting is called The Law of Assumption. Under the Law of Assumption, the designer simplifies his rules by assuming that the player will divine and comprehend the full meaning and breadth of such rules by knowing what they were supposed to have said in entirety. The designer, therefore, assumes that the player will make the same assumption as the designer did when designing the game. And the Law of Assumption *is* "Whatever Assumption You Leave To The Player, He Will Automatically Assume The Opposite!"

Of course, you cannot cover every conceivable happenstance in a game, which is why virtually every game, regardless of its level of excellence, has "errata." But you must make sure that you are not leaving the design/development work to the player. One of the classic examples of this is the "infamous" Supply Rule in John Hill's *Bar Lev,* a rule that left supply specifically to the players' imagination. Between two pliable players, no problem; between two homebred assassins, a very gritty evening! Another example of assumption would be a whole series of games that came out in the early 70's (from several companies) where it was not stated specifically whether a friendly unit would negate an enemy Zone of Control. There the player had to take a guess; he had a 50% chance of being right—poor odds where rules-writing is concerned.

Another important aspect of style in rules writing is the manner in which the rules and the game are presented to the reader/player. Most players, it must be remembered, are anxious to get into the game as quickly as possible. With simple games, there is no problem. But with a game such as *Terrible Swift Sword,* which contains many concepts and complexities, there is no way for the gamer to jump in without a bit of work. And this is where the Sequence of Play comes in.

The Sequence of Play is, by far, the single most important section of rules you will write. There is no danger of this statement being overemphasized, for the Sequence of Play is, as was explained earlier, the skeleton on which the game hangs. It is the barest, most simplistic, description on how to play the game. Granted, many games (mostly older, less-sophisticated, ones) use the simple Move/Fight sequence. But as games became more complex and as the state of the art expanded, the necessity for comprehensive Sequences of Play grew apace. A perfect example of this is the game *Conquistador. Conquistador* is a game in which each player has a great deal to do and keep track of. *When* he does these things and in *what order* becomes paramount to the smooth functioning and enjoyment of the game. Thus, the Sequence of Play becomes the focal point of the game, as players refer to it constantly to check on their activities and keep track of progress.

There is an additional importance for the Sequence of Play: it controls the intent of the game. For example, while most games follow the basic Move/Fight sequence, it was decided during the development of *Sixth Fleet* to *reverse* that time-honored method! Thus, in *Sixth Fleet,* players fight and *then* move! The difference was astonishing (so astonishing to some that they almost could not comprehend the game system, when all that had been changed was the sequence!). In tactical-level games, the Sequence of Play grows even larger in importance, for there units are performing functions the placement of which will greatly affect play.

A prime example of this latter case is *Terrible Swift Sword.* In that game, the designer purposely placed a Defensive Fire Phase *after* Offensive Movement to give the game a defensive bias. It enables the defense to *react* to the offensive player's moves, rather than *anticipate.* Such a decision must be made during the design stage of the game, although playtesting can reveal flaws and cause it to be changed. But changes in Sequence of Play can have spectacular effects on the game, so potential designers should be aware of the problems inherent in this area.

Most rules, for most games, are written in the same general format: an introduction, the sequence of play, the basic rules needed for play, and any optional rules or variants for extended play. Recently, however, Avalon Hill devised a different system of presenting rules, evident in their *Tobruk* and *Starship Trooper* games, whereby the player is given an introductory scenario as a complete, initial body of rules. This is followed by different scenarios in which the designer adds specific rules, making each succeeding scenario more complex than the previous, until, in the last scenario, the complete rules to the game have been presented. The idea is interesting, although it would seem to work only on multi-scenario (i.e., tactical-level) games. It is not sure whether this approach is advisable for amateur designers, as it takes a great deal of experience and discipline to produce such a body of rules. However, it is interesting to look at for possible ideas.

Unfortunately, one of the concepts of rules-writing is that you must write for the Lowest Common Denominator; i.e., every wargamer is not Albert Einstein. Happily, the average intellectual level of wargamers is quite high, so you do not have to stoop to phrases such as, "This is the game; the game is played with dice. . . . " But it is important to understand that for the game to be played, the rules must be understood. It is therefore recommended that designers group their rules into specific sections, subdividing these sections into smaller groups for cases within the major groups, as was demonstrated several pages earlier.

It would actually be a good idea to take this one step further and simplify the design/rules process even more. If the designer/writer can produce a simple outline of the rules that he wishes to cover and, during the course of design and/or writing the rules, note within that outline the various secondary rules that he wishes to put in (so as not to forget them), he will find life that much easier. Just a simple listing of 1 through whatever, encompassing the major sections to be covered with space for notes and subdivisions, would be a major time and effort saver.

Another item to consider is the terminology to be used. The designer must decide whether to use "layman's" phrases, or the official jargon of the hobby. In an appendix to this book is a listing of most of the terms accepted in wargaming today. Use these terms; they are recognized and used by almost all gamers and their usage will simplify and streamline your writing process.

The art of writing rules has come a long way since the early days of gaming, when a monster like *1914* proved virtually incomprehensible because of the problems with the rules. Yet the approaches to writing rules are as varied as the companies that produce games. In attempting to write rules, choose a style that is comfortable to you, one that you are familiar with. But remember, you want the players to play your game with as little fuss as possible, and you don't want your mailbox littered with rules questions. So be concise and be definitive. And above all, be readable.

PLAYTESTING

Playtesting is a luxury; but it is a necessary luxury. It is the only way for the designer/developer to tell whether what he has done works. More hours must be put into playtesting than into any other aspect of design work because, in the end, this is the single aspect of design work that proves out your results.

Playtesting means people. And, except for the large urban centers, people—especially wargame people—are not easy to come by. It is not recommended that designers do their own playtesting; they tend to play the game exactly in the way that they intended it to be played. They do not see the flaws, they fail to take advantage of loopholes, and they idealize their conception. In short, they are not critical. The ideal playtester is a well-known stranger, an idea that is as paradoxical in statement as it is to fulfill.

Ideally, a playtester should not be associated in any way with the designer. This is not to say that the designer should go around the streets of his local town soliciting strangers for an evening of relaxation (especially if the city is Salt Lake City). But the playtester should be someone who owes nothing to the designer, and vice versa. Moreover, the playtester should be well-versed in games. He should not only know how to *play* wargames, he should also be familiar with other designs and concepts, so that he has a level of comparison.

One of the reasons that SPI is located in New York City is the ready fount of playtesters the city provides. It is not unusual to see upwards of 25 people wandering through SPI's halls on Friday nights, waiting to test the new batch of games. Avalon Hill has a more formalized approach, using the services of Interest Group Baltimore, a highly competent but somewhat select group of gamers. Both companies use out-of-house testers, in addition to those whom they can keep under direct surveillance. The smaller companies, such as Battleline and GDW, usually have their coterie of testers upon whom they can rely.

And reliability is the most important trait in a tester. The designer/developer who knows his tester, who knows what reactions he has and what ideas he favors, is a lucky designer. The designer has to trust his playtester, for the tester is telling the designer what "he" thinks of the game. If that opinion is not valid, all playtesting is for naught.

Playtesters should be watched carefully. Remember, that they are not playing the game for fun (although that may be *their* intention). They are playing it to discern flaws, to comment incisively on the playability, and to make suggestions. A playtester who, after four hours of work, comments, "It's a good game, but too long . . . " is telling you nothing. Why is it good? And why is it too long? And what does he mean by too long? These are only superficial questions. Playtesters should write down any question that occurs to them during the course of the game. They should also make a note of any changes they make during the course of their play. Failure to do the latter can have some very strange effects.

When SPI's *Global War* was undergoing development, excitement was quite high at the company. The playtesters were unreservedly enthusiastic about the game, and they were quite enthusiastic about the system. Gamers who have played the final version of *GW* may find this somewhat amusing; but it is true. What had happened was that the developers failed to maintain any form of control—or even surveillance—of the testers. The testers were simply changing the rules as they went along (probably for the better) and not informing the developers. The developers assumed that the printed rules were the ones being played, and that was what went to the printer. Unfortunately the printed rules were not the played rules and the result was somewhat less than satisfactory from a design viewpoint.

Thus, the playtester is an extension of the developer. He must know what the developer is trying to do and how he is trying to do it. He must be able to play the game enough times so that both he and the developer can find out just what is and what is not working. In the early stages, he should know he has to play for design flaws, for the system. At the later stages, he plays for balance—or, as they say, to win. The tester who plays to win with a game in the 3rd week of development is less than useless, for he is only interested in himself. The good tester is working for you; he gets his enjoyment not only from playing but also from developing the game, albeit vicariously.

A word should be said here about "blindtesters." Blindtesters are testers who are usually sent a virtually finished copy of the game by mail to test in their own home, at their leisure. A blindtester is usually a person who can "nitpick", go through rules and spot minute flaws, as well as comment on the general tenor of the game. SPI and AH use some of the finest gamers in the country as blindtesters, gamers known to both companies as experts in both playing and testing. Blindtesting can be valuable, but it should not be used as a substitute for actual supervised testing, for only with supervised testers can you achieve meaningful results.

Playtesting is a labor of love. Testers are often unpaid and unsung (only recently have companies begun to list them in design credits). But testing can be fun, and it is important. Know your Testers and rely on their judgment. After all, they're playing your game much as would a stranger. What they are saying is what the potential buyer is going to tell you.

GAME DEVELOPMENT

Game development is probably the least understood and least practiced aspect of game publication. Basically, game development is nothing more than finishing a design. As noted previously, virtually anyone can design a game. Anyone can make up his own CRT, assign movement allowances and combat strengths. But all that information alone does not make a game, as numerous examples in past years can attest. The designer might have great ideas, but they do not mean anything unless they are developed. Unfortunately, game development is often given short shrift. This is partially because it is the one aspect of game publication, aside from art work, that is just plain old-fashioned work. In addition, development is a very difficult undertaking. In many ways it is like a science; but there is an element of art involved wherein developers can leave their personal touch on a game. It is also a frustrating phase, especially if the developer is different from the designer. The developer often feels that he is merely cleaning up the designer's blunders, and sometimes he is. However, there is often a wide gap between what the designer wants to do and what is required to make it work.

The Basics of Development

The first important element of the designer/developer relationship, assuming they are different people, is that the designer must be the one in charge. Someone has to be in charge of any operation and, in the end, it is the designer's creation that the developer is tinkering with. Initially, the developer will not know what goals the designer is trying to reach in the game. The only way to get around this is to develop procedures and methods which will cut down on this lack of information. One of the procedures that SPI has used is to have a fairly rigid structure for game development. The game will go through a series of stages, and at each stage something has to be done. For example, the developer has to have a complete set of rules. Initially, SPI made up the full rules after *all* of the development was done. However, problems arose from a lack of clarity in rules created at such a late stage. This could be prevented only by moving rules-writing forward in the process so that the development and playtesting would also be testing both the way the rules were expressed as well as the basic testing of the rule itself.

A developer also has to consider the audience the game will reach. There are certain players who play for blood. They like to study the game for a week before they engage in a "nice, friendly game." This type of player has to be considered to a great extent as they are the ones who look for loopholes. The developer must also think of the relative novice who may have trouble dealing with a set of rules which is aimed only at thwarting the loophole seeker. In effect, the designer must walk the thin line between rules which are so broadly written that the novice can easily understand them but which contain too many loopholes, and the other extreme—rules that are so dense that they can be understood only through repeated readings.

The developer must also be aware of the game's potential audience for another reason. The designer may well have ignored the potential audience in formulating his original design. The developer must then make the necessary changes to remedy this situation.

There are not many good game developers around. As more good developers appear, more good games will be seen, for it is not the designer or the design that makes a good published game; it is the developer. The designer will make the game come to life but it is the developer who transforms the game from a simple idea and collection of facts to a finished, polished product.

Part of the SPI method is to have the designer work hand in hand with the developer. He does not drop the design and walk away when the developer comes into the picture; he follows the game all the way through. A case might be made for developing the game twice! One person takes it through, and then another developer, starting pretty much at the beginning, develops it again. The second person would, in effect, be developing a finished game. It would be expensive, and it is beyond most companies' present capabilities; but the result in quality would probably be good.

Normally, the developer gets a map, in the sense of a playing surface, counters, and preliminary rules. These preliminary rules are only a structure which attempts to define what the game will be.

The first step the developer takes is to decide what he will do to continue the intent of the design. Starting with the map, it is essential to check the designer's sources. Find out where he got the map. This very basic step, if not done now, may lead to disasterous consequences later when the developer finds that he has built some rules to get around a terrain problem which later proves not to exist.

The second step is to determine the scale. Not knowing the scale the designer intended can yield some strange results in the development of the game. At some point the developer should list the various features which will go on the map, including town names, rivers, and any other geographical features. The best time to do this is when checking the map sources.

In considering the counter-mix, sometimes the developer receives a complete mix; but sometimes the mix is rather openended in that the counter-mix will be fixed by the scenarios. Whatever happens, at this point the developer must fix the size of the game being produced. Within what limits is he working, in terms of total game size and eventual market price? The number of counters is very important to the eventual published price of the game. If the developer finds the total counter allotment filled when he gets it from the designer, he should immediately recognize a potential trouble spot. Perhaps nothing will come of it but a full counter mix from the designer leaves no room for addition on the part of the developer. Furthermore, shaping and pruning the countermix to fit publication restrictions is an important—and difficult—process. It demands insight, skill, and a great deal of experience.

A subsidiary of this question is identifying what the counters represent. Are they regiments? Corps? Armies? This is important to the developer because he will have to determine how much information can be placed on the counter. The lower the level of the game—regiment and below—the greater the information which appears on the counter. This also will affect the complexity of the game.

The developer thus must know, at the outset, the level of complexity sought by the designer. Often a game has a way of taking off on its own, making an initially simple game complex. But the developer should not abet this tendency unless it can't be avoided. Usually the developer can expect too little information on the counter, rather than too much. This problem can carry over into playtesting, because nobody wants to spend the time to put down all of the data on the counters. This is fine as long as someone remembers that it must be added sooner or later, but the sooner, the better. A rule of thumb might be that the simpler a counter is—such as the "piece-of-change" counters of *World War I* with little more than the nationality and strength on it—the greater the indication of a simple game.

Having analyzed the counters, the developer should next analyze the preliminary rules. There are two points to clarify very early in development. First of all, what is the Player's role? What is he doing and whose historical position is he assuming? Is he going to be a one-role player, such as General McDowell at Bull Run, or is he going to be a multi-role player, á la *Solomons Campaign?* Are there different points where he has to be a ground commander, an air commander, a naval task force commander, an air strike commander? This all relates to the next important point: what is the sequence of play? A designer, having an over-all view of how the game progresses, will have certain things that he expects the player to do. He will have tested the game, probably by himself, to see that the game develops within a certain historical parameter. The developer's problem is to write rules to tell the other player when and where to put certain things and in what order. All of these things, if listed in the proper sequence of play, can allow the developer to formulate a coherent, complete set of rules, defining movement and combat rules. And of all rules, the sequence of play is *the* most important. It defines the structure of the game and acts as the skeleton on which the meat of the game will hang. The importance of the sequence of play, therefore, can never be underestimated!

The last initial item on development is asking the designer how he expects anyone to win. The designer may only have a vague notion, based upon his own feelings on the direction of the game. But having an idea of what direction the players are heading is important in determining the intermediate steps.

Having done this, the developer can now step into the actual development functions. Turning to the counter mix, he should begin to block out rules for every piece of information which appears on the counter.

Some changes may be indicated to the developer which are actually a basic design decision. The way the designer has arranged the grain of the map in most cases either facilitates or retards combat progression, except in the tactical games where one shoots through units. If the developer feels it is necessary to change the map, that is something to settle with the designer. Of course, there is nothing to prevent the developer from testing a change in the direction of the grain. Obviously this will make any suggestion he has on the final recommendation that much more valid.

The Mechanics of Development

Development is progressive. That is, it is a continuation and an evolution of design. Development is also creative. This creativity is manifested in the integration of test results into the rules and body of the game. Where the designer has to say "Should I include this or leave it out?" the developer has to answer the more simple but equally important question, "Is it good or bad?" He has to be able to point out what will work and what will not work and come up with a replacement for the latter. Development also incorporates new design elements placed in the game either by the developer or the designer or a combination of the two. Third, development is restrictive. This, in a way, is the very essence of development. It seeks to define the game limits and it also seeks to keep the game within certain cost figures.

Development requires continuity. The same person must be in charge from the start of development to the end. *The Destruction of Army Group Center* suffered from the fact that it went through two or three developers. An important technique in development is taking notes. A good developer should be able to document how he got from one point to another. This becomes critical when, in the later stages, he might come to a creative dead end. He has to be able to back-track up the chain of development to find the point where a change can conveniently be made. This can only be done with a good set of notes. Keeping notes also helps to explain the rationale behind a rule, which may be forgotten as development proceeds.

Quite often elements in the design may appear unusual, but if they are founded upon some historical basis, they should remain unless there is good reason to remove them. But in order to know whether something is historical, and when it is trivial, the developer has to be almost as well-versed in the period in question as the designer. In essence, the developer has to acquire the experience, knowledge and expertise to be able to challenge the designer and make changes on his own initiative.

Development involves very practical considerations. Taking *The American Civil War* as an example, the initial try at playing the game showed that the forces were too mobile. The solution was to enlarge the effect of command control. This was a rather sophisticated developmental decision: the developer first had to identify the problem and then come up with a viable solution, neither of which was necessarily obvious. Whereas on the original command control chart about forty percent of the Union Forces would be immobilized on turns one and two, and the immobility gradually reduced to about sixteen percent by the end of the war, under the revision the Union started with a *sixty* percent immobility rate. The South had a higher level of command control, which meant that it started out lower on the command control chart than the Union. Although the command control problem was still fairly severe at the end of the war, at any given time the North would have about twenty percent immobile and the South about sixteen percent. All of this was formulated as the developer watched the playtesters go through their paces.

The Year of the Rat was another game changed extensively during development. It originally was on a regimental scale. However, the campaign was an unusual one, and special rules had to be added to capture that flavor. The result was that the regimental level system was abandoned in favor of divisions. With divisions the number of pieces on the map could be reduced, although it was still necessary to keep brigades and specialized units of the ARVN because there were so many of them that they could not be ignored.

The element in *YOTR* which created the most problems and generated the most units was the use of limited intelligence. This was *the* key factor in the campaign, and it had to be included. The North Vietnamese had been marshalling by the border for quite a few months before they began their attack. The ARVN knew they were coming, but they did not know exactly where. Initial testing without this system showed that the ARVN *had* the strength to counter the North Vietnamese; in the real campaign they did not use this strength effectively. One can ascribe a certain amount of that failure to poor intelligence gathering by the ARVN high command, but that point can be carried only so far because the US was giving the ARVN a lot of good intelligence. The ARVN also realized that they had their backs against the wall and if they made too many mistakes, it was going to be all over. By this time the Americans were pulled back into a few enclaves and the North Vietnamese could simply sweep past the enclaves, beat the South Vietnamese, and take over. The idiocy factor wasn't critical among the ARVN; however, the simple fact was that they didn't know where the NVA was headed. And that is how the game ended up with "upside-down" counters.

Another problem was that the ARVN intelligence got better as the campaign went on. This led to removing the inverted counters one per turn as the game went on, as well as removing them when they were destroyed. A third problem in simulation was that the NVA was very successful initially and lost steam as time passed. The question was, why did this happen? It could have been better ARVN intelligence, which was doubtful, or an improved ARVN army, which was equally doubtful. Another possibility was the fact that the NVA was carrying its supplies with it and simply began to run out as the campaign went on. This last seems the best conclusion and the game tries to reflect it. In fact, the game, despite any other flaws it might have, has the ideal situation where one side has the initiative early, the game stalemates in the middle and the other side is on the offensive at the end.

Development often includes elements of integration simply because when a problem is found the basic historical element must now be reintegrated to produce a better result.

In *The American Civil War,* it became apparent, during development, that the supply system created by the designer was not viable. As a result, the whole system had to be rethought, and the present system of varying point values assigned. The victory conditions then became a matter of eliminating the number of supply points the Confederates had on one network of rail lines and navigable rivers. This means that the easiest way for the North to win is to march to the sea—out of the Cumberland Gap and into Charleston or Savannah. This cuts the southern resources in half, as well as destroying about fifteen percent of the resources enroute. A large portion of the South's production late in the war was concentrated in the Georgia area because the main production areas of the earlier part of the war, such as Tennessee and along the Mississippi, were lost. Oddly, as the game was developed, none of the playtesters saw any incentive to go down the Mississippi.

Development is, in effect, minor integration. It is an example of fine tuning. Although the *American Civil War* has been used as an example, that doesn't mean that it is a standard model. The solution which works in one game may not work in the next, and the developer has to be quick to spot this in order not to avoid spending wasted effort trying to get an unsuitable system working.

A person's skill as a developer is dependent, to a large degree, on his knowledge of history. This is not limited to the history that goes into the game but also the history of game systems. One can avoid a lot of mistakes by knowing that a certain method of handling a matter will have a given result. This is a direct result of having worked on many previous games.

A Special Word on Rules

It is hard to say just what effect the quality of the rules has on the eventual rating of a game. SPI has, in addition to its *Feedback,* a system called *Playback,* which is nothing more than a summary of player reactions to various elements of the game. *Feedback* is used to predict the probable receptiveness of a game; *Playback* measures how accurate the prediction was. One reason for doing the *Playback* is to obtain an item-by-item rating of each game. One can then see that some games get very low ratings because of the incompleteness of the rules.

SPI has found it a valuable exercise for the developer to review the rules after they have been set to type and pasted down. A photostat is made of the rules and the developer then puts the rules through a "strainer," since this is the first time the developer has seen the rules in something other than a manuscript format. Usually the manuscript is filled with pencil changes and deletions by the time it is set to type, and strange things have either crept into the rules or rules which were intended to be deleted weren't deleted, leaving utter confusion. The developer has a chance to correct these obvious errors at this stage. However, this is a step that the amateur companies often do not take with the obvious results that glaring errors remain in their rules.

In games, more so than with any other type of document (with the possible exception of legal documents), one word, one sentence, even a comma or a period can change the whole meaning of a rule. Accordingly, the rules have to be gone over quite thoroughly before they are sent off to the printer. Bear in mind, in citing the SPI situation, SPI sets its own type and *then* sends the materials off to the printers. Amateurs who cannot do their own setting will have to get a galley proof back if they want to take this step and this is an additional expense, since most printers will only correct their own errors for free. If the manuscript is faulty when it goes to the printer, that is not the printer's fault. So, for an amateur this step can only be performed cheaply by typing a clean draft of the rules as a "final draft" and going through the same steps SPI would go through on a photostat or a galley. A developer can spot more problems with the final rules than he can with the manuscript.

If a game is poorly developed and the problems with the rules are not spotted before publication, they will be reflected in the buyers' responses. Either they will complain and ask for clarifications, or they will stop playing the game and perhaps not buy any more of a given company's games. This is what is going to cause the demise of many new game companies. They will come out with games that people will simply not tolerate. Some of the early SPI games were atrociously developed. Speaking cynically, the best thing that happened to SPI was *Jutland, 1914, Anzio* and *Luftwaffe,* because these four games are really convoluted, each in its own way. As games, there were parts in them that were fun: *1914* if you liked slugging your way through France; *Jutland* if you had strong knees; *Luftwaffe* if you didn't mind losing to the Americans all of the time; *Anzio* if you didn't mind losing your eyesight and getting bloody fingertips from sorting counters. But the rules in each game were poorly developed, and the lessons learned from each were important.

Scenarios are a Special Problem

Doing scenarios increases development cost considerably. Notwithstanding contrary appearances, scenarios amount to almost a whole new game and have to be treated as such. In fact, a game with scenarios is more difficult than one without because the rules must be flexible enough to permit all of the scenarios to be played.

There are two kinds of scenarios. The developer is usually the one who creates them, though the designer may well furnish general suggestions if the situation obviously indicates scenarios. The first kind of scenario is the free-standing (or game-in-itself) scenario. This is common in SPI tactical games and requires a considerable investment in time and effort to develop. The other type is the "what if" scenario. This is just a variation on the basic game and is not as much of a problem to develop. This is the type found mostly in the strategic games.

One way around the cost of the free-standing scenario is to use a cross-breed between that and the "what if" scenario. This was used in *Desert War* and *Sniper!* This is done by simply mixing the situations and the order of battle. The developer then merely must de-bug the situations. For example, in a game like *Tank!,* with Shermans against Panthers, the Panthers have to have a very high panic level or they will just walk all over the Shermans. This is an historically valid solution (again, reflecting the need of the developer to know his history) because in 1944 the Germans on the Western Front were throwing in Panthers with untrained crews. The Americans and British in their Shermans were much more experienced, having up to six months experience. As a result, they shot up far more Panthers than the straight comparison of the vehicles would suggest as likely. This is a typical means of adjusting a scenario.

From a development point of view, there is really no such thing as an introductory version of any game. If there is an introductory game and a regular game, they will be two different games. There are very few game situations where a designer can simply have an introductory game by deleting a rule here and a rule there. It can be done to a point but requires an enormous investment of manpower, and once beyond that point the whole complexion of the game changes.

Development is where the frills are added. When someone is designing the game, the only thing he is really concerned with is getting the thing to move. A simple element such as using unit designations on counters is a developer function. The playtesters will not use unit designations unless the game requires a fixed set-up where designations are critical. They will simply mark combat and movement values on the counters and playtest. But, as already noted, the presence of unit designations is something the buyers want to see so the developer must be the one to check the order of battle and make sure that the counters have designations when they go to the printer.

It is common to blame a lot of the problems in games on inadequate playtesting. However, SPI's experience is that the real problem lies in a breakdown in communications. More often than not the playtesters either have not been adequately informed as to where the developer wants to go or they have located a problem area but the developer either is not informed of this discovery or fails to react to it.

If someone is willing to spend enough money, time and effort, he can produce an excellent game. Game development is the key. If someone can't develop games, he can't publish real games. He could probably publish something he could call a game, and he might even publish a game that actually worked; but unless the developer pays attention to his P's and Q's, he won't end up with a really good game.

It is critical that the developer have control of the game. Getting control means having a good, workman-like design at the onset, avoiding pit-falls, and having a set of rules in the very beginning which can easily be revised, rather than having to be fully rewritten.

Control also means that the developer has to keep track of the game. In *American Civil War* a situation arose concerning the leadership rules. There was a concept which was thrown out early in the process but somehow managed to stay in the rules. It was then carried forward until it was finally spotted again and at last fully eliminated. Without control that concept would have remained in the rules through publication and caused considerable confusion.

Development is the most crucial part of publishing a game. It doesn't matter whether the designer/publisher is a novice or experienced. It doesn't even matter whether the game is an exact reproduction of history or a virtual toy. If the development stage fails, the game will fail.

SECTION V:

Research: Sources and Materials

by David C. Isby

In order to be truly effective and valuable, a wargame requires a considerable amount of historical research. A game must not only be playable and enjoyable—it must also provide a valid and informative history lesson. Many people play the games to examine the historical events they depict and to experience the interactions between their own personalities and the elements of the historical battle or campaign. To allow this rewarding interplay, more is required of the designer than the correct unit designations and the right name-places on the map. This is where good research comes in, and good research needs good source material.

SOURCES: Archival and Personal Primary Source Materials and Secondary Source Material

Archival material is not readily accessible. Yet, for the serious scholar or researcher, this material fills a unique need. The researcher must ascertain what he is looking for and where it is kept. Sometimes the matter can be handled by correspondence or inter-library loan. Most often, however, a personal visit is required.

The archives are the home of many wonderful and useful documents. There are the original source materials, written at the time of the event or in the immediate light of it. Much valuable information that has never been fully extracted is contained within these documents. Thus, a good historian will be able to use these documents to his advantage, as they have not been affected by another's interpretation. They must be used with care, however, since many of them were written before the full facts of the event were available.

Such documents include such things as After Action Reports, which units complete after battles, and Intelligence Reports and Summaries, which show how a particular command level saw the situation at any one time. There are also Orders of Battle, Operations Reports, correspondence, orders, situation reports, staff and committee reports, and more of interest.

Personal primary source material is also most useful. Speaking to those who actually participated in the historical event in question can be most enlightening, provided that such information—as is the case with all historical information—is weighed and assessed critically. The good researcher must accept nothing on faith.

The main source, however, will have to be printed material: books, pamphlets, and magazines. There is an enormous amount of material on virtually any subject, even though the field of military history has been under-studied over the years. This chapter can hardly scratch the surface. What it will do is suggest possible "leads" for research, especially those sources that involve material vital for historical games that are not often available elsewhere, as well as suggesting some "basic" accounts that should give the reader a good overview of his situation and lead him on to other things.

More frustrating than where to find the information required for a game is where to find the books containing this information. No one library will have all of the possible sources. To be even partially successful, you will have to try the central library of a major city or university. Again, if the information you are looking for is not there, it can sometimes be obtained by inter-library loan, or, more usually, by going to where the source is. When you get to the library, make sure to look up all books under title, author, and all potential subjects before deciding it is not present. It also never hurts to ask the librarian on such matters. Some of the books you will be interested in will be "in print" at any given moment. Any decent bookshop should have a copy of *Books in Print* and *Paperback Books in Print*. Look through these under both title and author. If the book is listed, the bookstore should be able to order it for you. Used-book stores can be good for out-of-print books, but this requires luck more than anything else. Of the books in print, the researcher will find that all of the foreign-language books and many of the English language ones are published abroad. To obtain copies of these, a researcher might check the Yellow Pages of the nearest large city for a book importer, or consult a literary journal of the country whose book you are interested in. These literary publications often feature ads for mail-order booksellers.

Many of the traditional tools of the researcher are applicable in the field of historical games as well. Useful non-specialized bibliographies, indices, and abstracts include: *Historical Abstracts, International Bibliography of Historical Science, Bibliographic Index, Peace Research Abstracts Journal, British Humanities Index Writings on British History, A World Bibliography of Bibliographies,* and *The New York Times Index,* (good for very modern stuff).

Specialized bibliographies include Janet Ziegler's *World War II: Books in English,* useful but somewhat dated; Cyril Fall's *War Books,* useful for the First World War but more dated; Robin Higham's *A Guide to the Sources of British Military History* and *A Guide to the Sources of United States Military History;* C. E. Dornbusch's *Military Bibliography of the Civil War;* Allan Nevin's *Civil War Books: A Critical Bibliography;* Jacques de Launay's *The Two World Wars: A Selective Bibliography;* Bloomberg & Weller's *World War II and Its Origins;* Robert Albion's *Naval and Maritime History,* John Moran's *Creating a Legend* (about the USMC); C. E. Dornbusch's *Histories and Personal Narratives of the U.S. Army;* and *Soviet Military Power,* by the U.S. Army, reprinted by Greenwood Press. Note that one must keep in mind that bibliographies tend to "age" rapidly as new works appear.

Those wishing a general overview of the whole field of military history might try Montose's *War Through the Ages,* Fuller's *A Military History of the Western World,* Liddell Hart's *Strategy,* Zook and Higham's *A Short History of Modern Warfare,* Montgomery's *A History of Warfare,* Brodie's *From Crossbow to H-Bomb* (emphasizing weaponry), and Dupuy and Dupuy's *Encyclopedia of World Military History,* especially valuable for lesser-known conflicts. Harbottle's *Dictionary of Battles* is useful here.

Those looking for insights into the vast period of history preceding the Fall of Rome and normally known as "ancient history," might do well to look into such works as Frank Adcock's *The Greek and Macedonian Art of War,* Tarn's two volume work on *Alexander the Great,* J. F. C. Fuller's *The Generalship of Alexander the Great,* W. L. Roberts' *Naval Warfare Under Oars,* J. K. Anderson's *Military Theory and Practice in the Age of Xenophon,* W. L. Roberts' *Greek and Roman Naval Warfare,* Vegetius' *The Military Institution of the Romans,* Michael Grant's *The Army of the Caesars,* Michael Simkin's *The Roman Army,* and Hans Delbruck's *History of the Art of War—Antiquity.* For sources "closer to the bone," readers should try Herodotus, Arrian, and Livy, although they should be read in conjunction with the secondary sources for further explanation.

For the Middle Ages and Renaissance, one can do little better than the works of Sir Charles Oman: the two volume *A History of the Art of War in the Middle Ages, A.D. 378–1485* and a one volume *A History of the Art of War in the XVIth Century.* Supplementing these are such works as Terence Wise's *Medieval Warfare,* George Rush's *Renaissance Armies,* R. C. Smail's *Crusading Warfare,* D. F. Featherstone's *Bowmen of England,* and Beeker's *Warfare in Feudal Europe.*

The struggles of the Seventeenth Century are surveyed in Dodge's *Gustavus Adolphus* and C. V. Wedgewood's *The Thirty Years War.* The English Civil War has produced many good works, including H. C. B. Rodger's *Battles and Generals of the Civil Wars.* The opposing forces are studied in Peter Young's *The Royalist Army* and C. H. Firth's *Cromwell's Army.*

Eighteenth Century warfare was dominated by the personalities of the commanders. Therefore, we find David Chandler's two books, *The Art of Warfare in the Age of Marlborough* and *Marlborough as Military Commander* shed the most light on the War of the English Succession and the War of the Spanish Succession. Books by C. T. Atkinson and Winston S. Churchill on the Duke of Marlborough are also useful.

Frederick the Great dominated the Seven Years War, and Christopher Duffy's *The Army of Frederick the Great* shows what enabled him to do so. Julian Corbett's *England During the Seven Years War* emphasizes the strategic picture from the British angle.

In the Napoleonic era, David Chandler's *The Campaigns of Napoleon* and Esposito's and Elting's *A Military History and Atlas of the Napoleonic Wars* are the two most valuable single-volume treatments. There are many good biographies of Napoleon himself, but the best for military detail is probably Dodge's two-volume *Napoleon.* There are many worthwhile books on the specific battles and campaigns of this era, with many works written during the 19th and early 20th Centuries being especially well detailed. Any book by F. Loraine Petre or Jac Weller in this area, however, is bound to be useful.

The American Civil War was the most traumatic event in that country's history, and it has engendered a large amount of historical work. Good single-volume histories include those by Bruce Catton, Fletcher Pratt, and Dupuy and Dupuy. Analytic material may be found in Coggin's *Arms and Equipment of the Civil War,* Eisenschmil's *The Hidden Face of the Civil War,* Wagner's *Organization and Tactics,* Luvaas' *The Military Legacy of the Civil War,* Rodger's *Confederates and Federals at War,* among others, may be recommended out of the multitude of sources. Multi-volume histories include *Battles and Leaders of the Civil War,* and those by Shelby Foote, Allan Nevins, and Bruce Catton. Mark Boatner's *A Civil War Dictionary* is a good reference source.

Those looking for "harder" data on the American Civil War might try the *War of Rebellion: Official Records of the Union and Confederate Armies,* a grab-bag of reproduced original documents. There is a similarly titled set on naval operations. *The Official Illustrated War Record* and *Numbers and Losses in the Civil War in America* are good on numbers and strengths. There are also many individual volumes on the specific battles, such as Coddington's *The Gettysburg Campaign* and Wiley Sward's *Bloody April: The Battle of Shiloh.*

Other Nineteenth Century campaigns that are covered in good single volume accounts are: *The Franco-Prussian War* by Michael Howard; *The War Against Russia* (the Crimean War), by A. J. Barker; French Blake's *The Crimean War;* Phillip Warner's *The Crimean War—A Reassessment;* Jack Bauer's *The Mexican War,* Eversley Belfield's *The Boer War,* Gordon Craig's *Koniggratz* (Austro-Prussian War of 1866). All of these will give the researcher something substantial to chew on, and will have a good bibliography for further inquiry in most cases. Putting all of these wars in context is McElwee's *The Art of War: Waterloo to Mons.*

Overviews of Nineteenth Century colonial-type campaigns may be obtained from Brian Bond's *Victorian Military Campaigns,* Featherstone's *Colonial Small Wars 1837–1901,* and Farell's *Queen Victoria's Little Wars* for the British. American involvement may be seen from S. L. A. Marshall's *The Crimsoned Prairie* (about Indian fighting) and Dupuy and Bauer's *The Little Wars of America.* There also exist a fair amount of good works on specific campaigns of the period.

The First World War was the first total war, and so it spawned a huge crop of literature. Foremost among these are probably the Official Histories. While the British and others wrote official histories of many earlier wars, only the massive scale of the First World War mandated that the resources of the official machinery be required for a complete history. The best Official Histories are probably the British volumes, the most valuable being the scarlet-covered *Military Operations.* There are also series entitled *Naval Operations, The Merchant Navy, The War in the Air,* along with several statistical volumes and a set, *Orders of Battle,* from which much data may be extracted. Supplementing these are the Official Histories of the other Allied and Central powers nations: C. E. W. Bean's *Official History of Australia in the War; Canada in the Great War;* Ministre de Guerre, *Les Armees Francaise Dans la Grande Guerre; L'Escertio Italianie Nella Grande Guerra 1915–18; Official History of New Zealand's Effort in the Great War; The U.S Army in the World War; Official History of the Canadian Forces; Osterreich-Ungarns Letzer Krieg;* Bayerisches Kriegsarchiv, *Das Bayerisches Heer im der Grosser Weltkrieg;* and the German *Der Weltkrieg* by Der Reichsarchiv, supplemented by the many *feldgrau* covered monographs of *Schlacten Des Weltkriegs.* All of these are worthwhile and most helpful books for those who speak the language, although none of them are completely objective and many tend to be uneven, especially toward the end of the war.

While not an official history, the many volumes published by the Carnegie Endowment for International Peace contain much of interest, as does *The Times Documentary History of the War.*

To put this vast amount of information into context, the various one-volume histories are useful. The best is J. E. Edmond's *A Concise History of the First World War,* but those by B. H. Liddell Hart, Cyril Falls, J. F. C. Fuller, S. L. A. Marshall, and V. Esposito are also valuable. To single out a few from the many high quality non-official works on the First World War is most unfair, yet the various books of John Terraine, especially his *The Western Front,* Paul Fussell's *The Great War and Modern Memory,* and above all the various memoirs —those of Robert Graves, Seigfried Sassoon, Guy Chapman, and Edmund Blunden, being among the most literary, will have much for the researcher.

A word is perhaps in order here on unit histories. These are volumes that trace the activities of one specific unit and its personnel. These units range in size from companies to army groups. Many have value only in inducing nostalgia, while others are extremely well done studies of particular battles. The researcher will find unit histories most useful when dealing with tactical and operational level situations; but here too, their objectivity is never to be assumed, as most unit histories are quite naturally in sympathy with their subject. Therefore, their information should always be assessed with a critical eye (as should any information, primary or secondary) and, where possible, checked against the official history. Yet unit histories remain a valuable source, and many worthwhile books have used them as a source.

A most useful source for any research on 20th Century warfare is the various U.S. military Field Manuals and Technical Manuals (FM's and TM's). While most are concerned with incidentals, drills, and procedures, many contain most valuable information, especially the unclassified volumes of FM101, *The Staff Officer's Handbook,* and FM105, *The Maneuver Control Handbook*—the Army's wargame. Other volumes are useful on weapons, tactics, and organization. The main drawback is that these manuals are difficult to find, but some large libraries and universities which have (or had) an ROTC program have selections. Civilian publishers have also reprinted some manuals.

Those interested in post-industrial revolution tactical games will find themselves greatly concerned with the "software" aspects of tactics: cohesion, doctrine, and leadership, among other elements. The best book on this is S. L. A. Marshall's *Men Against Fire.* Supporting this is John Baynes' *Morale* and Morriss Janowitz's essay *Cohesion and Disintegration in the Wehrmacht,* which appeared in Public Opinion Quarterly in 1948, and more recently in Janowitz's book *Military Conflict.* Stouffer's multi-volume *The American Military Conflict* is full of hard data. On a more sociological level are Whyte's *Street Corner Society,* Tiger's *Men in Groups,* and Lord Moran's *The Anatomy of Courage.*

Pre-1914 tactics can be studied in Wagner's *Organization and Tactics,* Balck's two-volume *Tactics,* Eames' *The Rifle in War,* and other works. First World War tactics are covered in Balck's *The Development of Tactics During the War* and the excellent *Infantry in Battle,* which provides valuable insights for any period. First World War tank tactics are outlined in Fuller's book on the subject and Jones' *The Fighting Tanks.*

Second World War tactics are surprisingly poorly documented. Jac Weller's *Weapons and Tactics* has some value here, but there is a large amount of material to be dug out of sources such as the appendices of British official histories, memoirs, and other sources. Any book worth its salt on North Africa, for example, will explain how German tactical methods enabled them to gain many of their victories. The tactical "textbooks" of the German, Soviet, Italian, and Japanese Armies can be found in the U.S. Army *Handbooks* on these armies in the National Archives. The same applies to armored tactics, although Icks' *Famous Tank Battles* and several books by Richard Ogorkiewicz are good here. Twentieth Century artillery tactics are covered in Shelford Bidwell's *Gunners at War* and books by Fairfax Downey.

In dealing with modern tactics, Weller's *Fire and Movement* and *Weapons and Tactics* are, though dated, of some use. As with the Second World War, data may be culled from other sources, with many of the earlier ones still being valid. Here, as throughout the field of tactics, journal articles are especially valuable.

The Second World War also spawned a crop of official histories. The United States is perhaps the best represented with S. E. Morrison's well-written, fifteen volume *U.S. Naval Operations in World War Two.* Less well done, but still useful, is Craven and Cate's *Army Air Force in World War II,* some six volumes. The *U.S. Army in World War II* is some 83 volumes and is still unfinished. As might be expected, there are many different authors and quality varies, but it is usually quite high. *The History of U.S. Marine Corps Operations in World War II* is that service's official history. In addition to these, there are large numbers of monographs, unit histories, and reports that are full of useful data.

The British United Kingdom Military Series, edited by J. R. M. Butler, is less successful, but still possesses much useful data, especially in the appendices. Sub-series include *The Campaign in Norway* by T. K. Derry, *The Air Defense of Great Britain* by Basil Collier, *The Campaign in France and Flanders 1940* and *Victory Campaigns* by L. F. Ellis, *The War at Sea* by S. W. Roskill, *The Mediteranean and Middle East* by I. S. O. Playfair and others, *The War Against Japan* by S. W. Kirby, *Grand Strategy* by numerous authors, *The Strategic Air Offensive Against Germany* by Webster and Frankland, *British Foreign Policy in the Second World War* by I. L. Woodward, *The S.O.E. in France,* and, most valuable to the researcher, H. F. Joslen's *Orders of Battle.* There is also Hilary and Saunder's three-volume history of the R.A.F.

The British Commonwealth produced a number of high quality, yet generally quite nationalistic official histories. These include: *The Indian Armed Forces in World War II; Official History of New Zealand in the Second World War; South African Forces in World War II; Australia in the War of 1939–45;* and the *Official History of the Canadian Army.*

The German Armed Forces of the Second World War produced no official history, but the vast war effort of Nazi Germany has been fairly well studied. Georg Tessin's multi- (twelve plus) volume *Verbande und Truppen der Deutschen Wehrmacht und Waffen-SS* contains huge amounts of data on unit composition and orders of battle, although a fair amount of digging is required. The same can be said for *Der Kriegestagebuch der OKW,* another multi-volume set whose specialty is order of battles. A little patient work will show the disposition of most of the German army at any given moment. Two other valuable multi-volume works are Wolf Keilig's *Das Deutsche Heer* and *Das Heer 1933–1945* by Burkhart Muller-Hillebrand.

More order of battle data can be found in *The German Order of Battle, 1944,* published by Hippocrene, but somewhat incomplete, and in the *German Order of Battle, 1945,* done by the Army and found in some libraries and in the National Archives. The book published by Arco and Ian Allan, *The German Army Handbook,* is useful but also incomplete in parts. It should not be confused with the excellent U.S. Army *Handbook on German Military Forces, 1944 ed.* This is also available in some large libraries and in the National Archives, although a hatcheted version has been reprinted. F. M. von Senger und Etterlin's *Die Deutschen Panzer, Taschenbusch Der Panzer, Der Deutschen Geschutz,* and *Der Panzer-Grenadiere* are useful, as are the 50-plus monographs that make up the *Die Wehrmacht in Kampf* series.

Luftwaffe operations are covered in works by Cajus Bekker, John Killen, Alfred Price, among others. The best single-volume account is *The Rise and Fall of the German Air Force,* published by the Air Ministry after the war but since reprinted. Arno Press has also published a large series of monographs on Luftwaffe operations. Describing the air war over Germany and its interface with the rest of the German war effort is the multi-volume *U.S. Strategic Bombing Survey - Germany.*

The Soviet side of the war in the East has been less than adequately documented. The Soviet official history, *Istoriya Velikoy Otchestvennoy Voyny Sovetskoyo Soyuze* (six volumes, also in an edited East German translation, *Geschichte des Grossen Vaterlandischen Kriegs der Sovjet Union)* has some data, but suffers from "socialist realism" history; as does *The Soviet Air Force in World War II,* edited by Ray Wagner, the English translation of the Soviet Air Force's official history. Here, more than in other fields, one will have to turn to original sources, with TM30–40 *Handbook on USSR Military Forces,* and various captured German documents, especially *Der Kriegsgliederung der SU Heer.* Also useful (and available at many university libraries that have—or had—Army ROTC) is the U.S. Army's German Reports Series of monographs on the Eastern front, which contain large amounts of data not available elsewhere. The USAF Historical Studies/German Air Force Monograph Project is similar for air matters.

Japan's war effort has been documented by the U.S. Strategic Bombing survey, which delved into much of the entire war, not just the B-29's. Hayashi and Coox's *Ko-Gun* is a good one-volume account of the Imperial Army, while Okumiya's *Zero* does the same for naval aviation. Most useful are two scarce wartime publications: TM30–480, *The Handbook of Japanese Military Forces* and *Japanese Order of Battle.* As with the other War Department wartime publications, the accuracy is surprisingly good, but cross-checking with post-war sources were possible will prevent some clanging errors. *The Reports of General MacArthur* contain considerable information on the Japanese, as do most U.S. official sources.

Distressingly little has appeared on the Italians in any language except their own, although they do have an official history. The TM30–420, *Handbook on Italian Military Forces* thus becomes useful, but must be used with more caution than usual as it is an "early war" project.

Other useful sources for the Second World War include the *Geschichte des Zweiten Weltkreig,* later editions of the *League of Nations World Armaments Yearbook, The Army almanac (much useful OB data for the U.S. Army),* J. A. Huston's *The Sinews of War* (good, along with the U.S. official history volumes, for logistics), and the multi-volume *Eizelschrifter zur Militarisen Geschichte* series of monographs.

Again, the amount of quality literature on the Second World War cannot be dug into here, but those wishing to put their information into context or who are looking for a place to begin should try the one-volume histories by Basil Collier, J. F. C. Fuller, Henri Michel, Basil Liddell Hart, Peter Young, A. J. P. Taylor, and Hanson Baldwin.

For the Korean War, the U.S. Army, Navy, Marines, and Air Force all wrote official histories which are useful. Matthew Ridgway's *The Korean War,* David Rees' *The Limited War,* and Joseph Collins' *War in Peacetime* are good one-volume accounts, while the U.S. Army's Combat Actions in Korea gives the feel of the tactical situation. The official histories are James Field's *U.S. Naval Operations: Korea;* Robert Futrell's *The U.S. Air Force in Korea;* and the multi-volume *The U.S. Army in the Korean War* and *U.S. Marine Operations in Korea.*

For naval warfare, Michael Lewis' *The Navy of Britain* and Robison's *A History of Naval Tactics* are good analytical accounts of naval warfare under sail and after, while the most complete narratives are William Laird Clowe's *The Royal Navy* and Joseph Allen's *Battles of the British Navy.* Also useful are Michael Sanderson's *Sea Battles: A Reference Guide* and John Southworth's four-volume *War at Sea.* H. Wilson's *Ironclads in Action* and *Battleships in Action,* Reginald Custance's *The Ship of the Line in Battle,* Arthur Marder's *From Dreadnought to Scapa Flow* and *The Anatomy of British Seapower,* along with the various operational histories, official and non-official, and books on the ships and aircraft which carry the story over the last hundred years.

In dealing with modern military situations, for general data as to the strength of the world's armed forces, the Institute of Strategic Study's *Military Balance* is a good annual source. Supplementing it are the various editions of Dupuy's *Almanac of World Military Power.* The strengths of the world's navies may be found in *Jane's Fighting Ships, Flottes de Combat,* and Weyer's *Warships of the World.* Air Forces change very rapidly. The most recent source specifically on the aircraft is Norman Polamar's *World Combat Aircraft Directory.* Supplementing this, although slightly more dated, is N.Z. Krivinyi's *World Military Aviation.* There are many works on this topic, and any book by William Green in this field is generally good.

Armies are difficult to find specialized data on. Otto Von Pivka's *The Armies of Europe Today* and the three-volume German-language *Taschenbuch der Landstreitkrafte,* published by J. F. Lehmanns Verlag, provide a great deal of information on European armed forces and the great powers. For others, try the *Military Balance, Almanac of World Military Power,* or dig through magazines, although specialized studies do exist.

Data on "hardware" is, as always, easier to find. The three annual *Jane's* volumes—*Jane's Fighting Ships, Jane's World Aircraft, Jane's World Weapons Systems*—are the standard sources. Supplementing these, for the land weapons, are C. F. Foss's *Armored Fighting Vehicles of the World* and *Artillery of the World.* Brassey's *Infantry Weapons of the World* is good on its subject. In addition to the previously mentioned sources on aircraft, the annual *Observer's Book ofAircraft* is useful.

Maps will be one of the researcher's main problems. Many books are well-mapped in themselves. Others are not. Esposito's two-volume *West Point Atlas of American Wars* is the most useful source for maps, although the accuracy should be checked for maps in Volume 1. There are any number of atlases, both contemporary and historical, available, and these should be in any library. The U.S. Army topographic Command, Army Mapping Service, and Defense Mapping Service all produce many maps worldwide. These are sometimes available at large libraries, especially those with government documents collections. There are also "commercial" maps by publishers such as Hammond and Rand McNally that can be obtained through usual ways. The U.S. Coast and Geodetic Survey and U.S Army Corps of Engineers provide maps of the U.S. The British *Ordinance Survey* maps do the same for that country, and several other nations have similar services.

Magazines

A large amount of material not available elsewhere can only be found through research in magazines. Many libraries will have back issues files of these magazines running back many years, and these can be used for research. This list is aimed at contemporary magazines, however, and subscription data are included, along with a brief description of the use and purpose of the magazine.

One especially useful source for the researcher in this field is *Strategy & Tactics.* While it may seem somewhat immodest, it remains that *Strategy & Tactics* is one of the best and most widely circulated of any military history magazine in English.

PRIMARILY CONTEMPORARY

Aviation Week & Space Technology. $37.00 1221 Avenue of the Americas, NYC, 10020. Weekly. Much up-to-the-moment data on the latest in defense hardware. Emphasis on air material, but much good data in related areas as well.

International Defense Review. $24.00 86 Avenue Louis-Casai, Box 162, 1216 Cointrin, Geneva, Switzerland. Much good information on modern equipment, with a good deal of material not readily available elsewhere.

Infantry. $6.50. Bimonthly. Box 2005, Ft. Benning, GA, 31905. Modern infantry tactics, weapons, and organization are all well covered.

Orbis. $10.00. 3508 Market St., Phila. PA, 19104. quarterly. The most useful political science journal for the military researcher, this magazine contains many articles on interest on current strategic problems.

Armed Forces Journal. 1414 22nd St., NW, Washington, D.C., 20037. $12.00. Run by and for the "military industrial complex", this magazine still has material of interest to those looking for information on modern weapons.

CONTEMPORARY AND HISTORICAL (CONTEMPORARY PREDOMINATING)

Armor. $6.50. Bimonthly. Box O, Ft. Knox, KY, 40121. Mainly concerned with the organization, weapons, and tactics or armor and related fields, and it is valuable for those interested in the field. There is also some armor-related history.

U.S. Naval Institute Proceedings. $12.50. Annapolis. MD, 21402. Mainly concerned with sea power and its role in today's geopolitical situation, this most useful magazine also features articles on different facets of naval history.

Armies and Weapons. $15.00. Bimonthly. Sky Books Int'l, 521 5th Av, NYC, 10017. Despite occasional errors and translated English, this magazine has many useful articles on both contemporary and historical military situations. Its sister publication *Aviation & Marine* (identical data) does the same for naval and air matters.

Military Review, USACGSC, Ft. Leavenworth, KS, 66027. $9.00. The U.S. Army's professional journal contains much of interest, with a good deal of material on strategy, operations, foreign armies, and some history.

Army, 1529 18th St., N.W., Wash. D.C., 20036. $12.00. While aimed at the serving Army officer, it does contain many articles of value, both contemporary and historical.

Air Force, $10.00, 1750 Pennsylvania Av. NW, Washington, D.C., 20006. The current military situation as seen by the world's most powerful force; useful for air warfare material.

Journal of the Royal United Services Institution, $15.00. Whitehall, London, SW1, England. Quarterly. Scholarly journal featuring analysis of contemporary military problems, both strategic and tactical. Its pages have been the home of generations of military intellectuals.

HISTORICAL AND CONTEMPORARY (HISTORICAL PREDOMINATING)

Military Affairs, $10.00. KSU, Manhattan, KS, 66506, quarterly. This contains scholarly articles on many topics not treated elsewhere.

AFV-G2. $12.50. P.O. Box 820, La Puente, CA, 91747. Much information on armor, including organization, tactics, and weapon and vehicle performance.

Warship International, $8.00, 1729 Lois Ct., Toledo, OH, 43613. quarterly. Contains much in-depth information on warships. While much is very esoteric, it contains much of use for the researcher in naval affairs.

PRIMARILY HISTORICAL

Savage & Soldier, $5.00, Ian Cohen, ed., 12426 Kingcross, Ln, St. Louis, MO, 63141. Quarterly. An excellent journal dealing with colonial warfare, 1850–1918, especially as it relates to wargames.

Soldier of The Queen, $9.00, 18 Tudor Ct., Parks Rd, Teddington, Middx., England. Quarterly. Deals with the British Army of the 19th century and its campaigns.

Journal of the Society for Army Historical Research. $10.50. The Library, Old War Officle Bldg., Whitehall, London, SW1, England. Devoted to the British Army and contains much otherwise obscure information.

The Civil War Times Illustrated, $10.00 206 Hanover St., Gettysburg, PA 17325. Articles on the American Civil War.

All magazines are published monthly unless otherwise noted. Subscription costs and addresses are correct as of June, 1976, but these do change and it might be wise to enquire first.

Following are some bookshops that specialize in books likely to be of interest to the researcher and unavailable elsewhere. It is, of course, a highly abbreviated and inconsistent list: Beaumont Aviation Literature, 656 Holloway Rd., London, N19 3PD, England (specialist in aviation); The Bohemia Bookshop, 116 Bohemia Road, St-Leonards-on-Sea, Sussex, England; Bivouac Books, 104 Kilburn Square, London, NW6 6PS, England; The Woodford Bookshop, 33 Victoria Road, S. Woodford, London, E18 1 LJ England (specialist in British subjects and First World War); S. Carwin, Box 803, Tarzana, CA, 91356; Antheil, 2177 Isabelle Ct., N. Bellmore, NY, 11710 (naval); The Compleat Strategist, 11 E. 33rd St., NYC, 10016; The Soldier Shop, 1013 Madison Av., NYC, 10021; Armed Forces Book CO., 21 Dahlia Ln, Deer Park, NY, 11729, Sky Books International, 48 E. 50th St, NYC, 10022; Adler's Foreign Books, 162 Fifth Av., NYC, NY, 10011 (foreign, non-English language books), M.S. Rosenberg Inc, 100 W. 72nd St, NYC, 10023 (German language books). Hillcrest Publications, Box 395, McDonald, OH, 44437 (reprinted Army manuals).

In writing to any of these bookshops, be sure to enclose a stamped, self-addressed envelope, and, in the case of foreign bookstores, send along two International Postal Reply Coupons which are available at your local Post Office.

The above information can hardly pretend to have been a definitive bibliography. Rather, it has been a ramble through what the author (who has researched many games and articles for publication) thinks would be helpful to the researcher coming rather "cold" into the field.

Ignored to the greatest extent are non-published sources and accounts of specific battles and campaigns in favor of the official histories and analytical accounts. To offset this, good one-volume histories on as many subjects as possible have been recommended, as these, along with the various bibliographies and the card catalogue of your local large library, can provide a start for research. While neglecting such useful tools as books on weapons and similar hardware and unit histories, both of which the researcher will certainly have occasion to use, the list has tried to point out the more obscure sources that contain most of the useful information.

SECTION VI:

The Business of Wargaming

The Economic Realities of Wargame Manufacturing and Marketing

by James F. Dunnigan

Publishing has been viewed traditionally as a "gentlemen's profession" that one did not engage in for love of money. The concern of the publisher has been to produce the best possible publications. Yet, despite this devotion to quality, the facts of business dictate that no publisher can long endure without at least breaking financially even. Those publishers who show profits almost invariably survive the longest, a phenomenon that is attributable in large measure to the fact that they have the capacity to weather both external and self-inflicted emergencies.

Ultimately, both survival and its prerequisite, breakeven, depend on whether or not a sufficient number of people buy a publisher's product—the product, in this case, being games. And, as if to re-confirm the publisher's traditional image, the most crucial aspect of a game's salability is its quality. It is true that, in the business of game publishing, retail distribution is dominated by *impulse buyers*—non-gamers on whom the historical and technical quality of a wargame may well be lost. But it is also true that no game publisher can rely solely on the magnitude of his retail distribution network, no matter how extensive it is, to sustain his business. It is ultimately the *quality* of his entire line of games that determines the degree of his success or the magnitude of his failure. While gamers, who are unlike impulse buyers in that they *do* discern and compare the quality of games, may have legitimate complaints about particular games, they will ensure or prohibit a publisher's survival on the basis of his line's *overall* quality.

THE BIG TWO COMPANIES

Two major wargame publishers—SPI and Avalon Hill—account for a combined total of 85 to 90 percent of the wargame business. The means for calculating this kind of figure offers direct insight into how wargame companies perform as businesses.

WARGAMING MARKET SHARES
Between SPI, Avalon Hill, and All Other Wargame Publishers

	Percentage of Total by Publisher		
Type of Share	**SPI**	**AH**	**Other**
1. Units Sold	50	35	15
2. Bought by Wargamers	56	26	18
3. As above, less 120,000 S&T	46	32	22
4. Cash Spent by Gamers	43	43	14
5. Cash Received by Companies	49	34	17

Note each line of figures equals 100 percent.

Based on information received directly from publishers, or in the case of Avalon Hill, from corporate statements, there are three ways to calculate wargame sales: according to units sold (individual copies of games); according to cash spent by gamers; and according to cash actually received by the game publisher.

SPI currently sells some 450,000 war games a year, Avalon Hill some 320,000 and all other publishers combined, no more than 140,-000. As shown in the Market Share table, these figures represent 50, 35, and 15 percent of the market, respectively. But a comparison based strictly upon units sold creates some difficulties in defining—and assessing accurately—the wargame market. For instance, 90 percent of Avalon Hill's wargames are sold in retail stores. As far as can be determined, between 50 and 70 percent of these sales are made to non-gamers, many of whom will never actually play the game nor buy another ever again. It's clear then that there are at least two classes of wargame sales: those made to the gamer-hobbyist and those made to the impulse buyer. Approximately 83 percent of SPI sales, 55 percent of Avalon Hill sales, and 95 percent of other wargame company sales are made to the hobbyist-wargamer. The second line of the Market Share table reflects these sales.

If one knocks out 120,000 *S&T* subscription games, from SPI's share, the result is the third line of the table. This is a somewhat dubious adjustment since two-thirds of the subscribers play the subscription game that they receive—and the majority of subscribers buy S&T *because* it contains a game. All these methods of calculating the shares of the wargaming market are necessarily imprecise—not only because of the different mix of marketing environments that the various companies operate in, but also because of the availability of data.

SPI, in its two magazines, publishes just about everything anyone could want to know about its business and financial operations. So much in fact, that some of our readers have asked us to stop! Also, virtually all of SPI's business is wargame-centered. By contrast, Avalon Hill does not freely make available its financial data and sales figures. This is not to imply that AH is doing something "tricky" behind our collective backs—it's simply that Avalon Hill is wholly owned by a larger corporation the shares of which are publicly traded on the stock market and such corporations are traditionally very secretive about what they consider "inside" information. The information used here is derived from public reports which AH's owning corporation is required to file with the SEC and interpolations made from other legally required business statements. To be fair, it should be said that Avalon Hill is not purely a wargame company: approximately 25 percent of their business consists of the sale of sports games, word games, abstract games and the like. This portion of their business is not considered in the information discussed in this section. Since they acquired 3M's adult game line in early '76 (followed later by the acquisition of SI's sports game line) the percentage of Avalon Hill's business not relevant to wargaming can be expected to rise. In fact, Avalon Hill shows many signs of wishing to place its emphasis on its non-wargame business. Estimates of Avalon Hill's performance have routinely been published annually in *S&T* Magazine. SPI has duly solicited figures from Avalon Hill. The response has been, to date, that AH will neither confirm nor deny anything relevant to their sales. The figures on Avalon Hill are therefore approximate—but every effort has been made to make them as fair and as accurate as possible.

To continue our examination, both SPI and Avalon Hill attract about the same number of dollars from the gamer—but SPI actually receives a greater share of these dollars (see the last two lines of the table). This is because a greater fraction of Avalon Hill sales are made via retail distribution and, therefore, a greater amount of money is eroded by middleman profit taking (i.e., store owners and distributors).

In summary, as of approximately mid-77, SPI has a market share of 43 to 56 percent; Avalon Hill has 26 to 43 percent; and the smaller publishers together have 14 to 22 percent. The preceding analysis and that which is to follow should all be considered in the context of the scale of the wargaming "industry", which though vigorous and fascinating, is nevertheless pretty small compared to other branches of publishing and hobby industries.

Wargame Unit Sales: 1964–1976

Year	Avalon Hill	SPI	Other	Total
64	62,000	0	100	62,100
65	65,000	0	300	65,300
66	66,000	0	420	66,420
67	85,000	0	760	85,760
68	92,000	0	11,500	103,500
69	96,000	2,500	3,200	101,700
70	115,000	11,500	2,900	129,400
71	155,000	27,900	2,900	185,000
72	183,000	151,000	3,280	337,280
73	197,000	242,000	12,000	451,000
74	190,000	283,000	120,000	593,000
75	237,000	406,000	100,000	743,000
76	296,000	420,000	125,000	841,000

The sales figures represent only "wargames" (conflict simulation games) as typified by the sort of games first published by Avalon Hill and then SPI and others. Left out are such games as *Risk, Stratego* and *Diplomacy. Diplomacy* probably should be included but isn't because sales figures are impossible to come by. (One estimate is that *Diplomacy* sells some 15,000 copies a year, but there are any number of problems with that figure.) Also left out are sales of miniatures equipment, mainly because miniatures equipment is not sold as complete "games" but rather as parts of games. We estimate that for every dollar spent on board wargames, some 30–40¢ is spent on miniatures equipment (how's that for precision).

In the beginning there was Avalon Hill, and only Avalon Hill. They had been selling the games since 1958, but the sales figures for the pre-1964 period are unavailable. It is said that more games were sold in the early sixties than in the late sixties. It's anyone's guess. In 1964 Avalon Hill went bust and was re-organized that same year. Up till then AH had no competition, except for a few private games published and sold in miniscule quantities. Starting in 1968, the Gamescience Corporation came out with a number of games but ran into marketing and managerial problems which soon forced it from the scene. SPI got started in '69 and two years later began advertising on a large scale. About this time Avalon Hill began publishing more (and better) games (*PanzerBlitz, Anzio, France-'40, Luftwaffe,* etc). The combination apparently got gaming going on a large scale. There began to appear more smaller publishers. These folks usually (but not always) emulated SPI's nonmounted board, non-box approach. By 1974 a number of smaller publishers really got into high gear. The next year a number also collapsed while still more joined in. At this point Avalon Hill came out with still more games each year, and sales went even higher. SPI cut back on game publication in 1974 and then really kicked out the jams in '75 (over 40 games.) SPI changed again in '76 by coming out with a large number of much more expensive games. And so it goes.

SPI

SPI (or Simulations Publications, Inc.) could best be described as a company run like a game. SPI was founded by gamers and continues to be run by gamers. For better or worse, the company was set up using the same techniques employed in designing wargames, and ongoing operations are conducted much as one would play the "SPI Game". This means the management of the company is very numbers oriented. Everything, including the people working there, is constantly rated and evaluated: there is a lot of long range (5+ years) planning and mathematical modeling and projected-versus-actual results are constantly compared and adjustments made. The Feedback in the magazines is but one application of this approach. Despite all of this seemingly organized and efficient busywork, many decisions at SPI are still made either by intuition or "what the hell, let's try it". However, the numbers are often dragged out to prevent bloodshed should a dispute arise. Surprisingly—or not so—this aspect of the system works quite well.

SPI began as a direct mail operation, mainly because it required too much expertise and money to quickly get into the Avalon Hill approach (through the toy distribution system). SPI only got into the retail end when they discovered, through their feedback, that some 50 percent of their potential customers did not, and would not, buy by mail. Retail marketing also gave the SPI staff another chance to put the company out of business and, at the same time, gave them one more opportunity to irritate Avalon Hill (which is pursued, without malice, as a favorite pastime at SPI).

SPI publishes so many games mainly because there are so many to be published. They all sell (some, admittedly, better than others), and most of them are liked by most of the people who buy them. SPI tends to put a lot of emphasis on quality in game rules but, because of the large number of games they publish (30 or so a year), they often give the impression of "cranking" the product out. SPI thus tends to draw lots of flak, much as Avalon Hill did before they lost their near-monopoly position in the early 1970's. SPI *is* a bigger target and does have its share of flaws, but SPI also goes out of its way to find out what its customers want through its feedback system. Despite its relatively small size (as businesses go) SPI has the largest in-house group of game Research and Development people (thirteen, full-time) in the world. It publishes more wargames each year than the next three largest wargame companies combined. It is difficult to discuss SPI in comparison to anything else because there isn't anything else quite like it. The people at SPI tend to see themselves as magazine/book publishers. But even that view is stretching the point. Add to this the fact that few of the people managing SPI had any previous business experience or training and you have the makings of one very strange organization.

Avalon Hill

Avalon Hill (AH) is the most commercially-oriented of the wargame companies. It is also the oldest (since 1958) and most profitable (13% net, before taxes, for the year ending 4/30/76, compared to 6% for SPI and unknown amounts for other publishers). Avalon Hill has survived for so long mainly because it was always run in a businesslike manner. It also had a monopoly position in the wargame field for the first ten years of its existence. But Avalon Hill also had the burden of leading the push into uncharted (and rather dangerous) territory. This background explains much about why AH does things the way it does.

Avalon Hill was, like SPI, founded by gamers. But in 1964 the original management went broke and control of the company was taken over by non-gamer businessmen. This led to a period of extreme conservatism at AH, ending only when, (ironically enough) the management of AH's parent company again took an active interest in AH's management. This came as a result of a number of years of no-growth at AH while, at the same time, SPI was growing rapidly to its present size. Starting in the early 1970's AH began publishing more games (six or more a year) and selling more agressively by mail. As simple as these moves were, they entailed considerable internal changes.

The "new" AH required considerable readjustment in their traditional attitudes towards the wargame business. Previously, the people running Avalon Hill believed that to publish more than one or two new games a year would *saturate* the market. This may sound ridiculous today, but, between 1964 and the early 1970's, this was The Word. And there was nothing to contradict it. Needless to say, this indicated that AH was not experiment oriented. In addition they were considered to be "tight" with their dollars, which was one reason why they used independent designers for those games they did publish.

Once the decision was made, in the early 1970's, to increase the publication rate more people were added to their staff. (Up until then there was only *one* full-time employee involved in game R&D at Avalon Hill, and he only worked on games part of the time! A few part-timers were occasionally used, especially during the summer.) The new employees added many new ideas, as well an an impetus for further innovation. Despite all of this, Avalon Hill still remains more commercially-oriented than all the other wargame publishers. Given AH's continued dependence on the more commercial retail market, this orientation is not inappropriate nor does it appear to hurt them. Indeed, some of AH's older "trade customs" have been turned into distinct advantages under the new management approach. Their practice of using independent designers on a royalty basis has allowed them to publish a more eclectic selection of games than any other publisher. The rules are often a bit ragged (coming from so many different sources, and being edited by essentially new staff), but their customers don't seem to mind.

Recently, however, Avalon Hill has undertaken a project which may eventually have a profound effect on their wargame publishing activities. In early 1976 they bought the 3M adult game line. As usual, no detailed explanation of their actions was given. Those that will talk at AH say it was done to obtain the wider range of outlets 3M had developed. On the other hand, AH kept a number of 3M's fifty adult games and began energetically selling them along with their own adult games and wargames. Ever since AH's beginnings in the late 1950's they have been trying to orient the company more towards general adult games rather than pure wargames, although wargames have consistently been their bread and butter. All previous attempts to get the company primarily into adult games have failed; only wargames have worked consistently for Avalon Hill. Whether or not this latest venture into adult games will succeed is anyone's guess. What impact this will have on Avalon Hill's wargame publishing activities is also unsure. Time will tell.

The "Other" Wargame Publishers

Both Avalon Hill and SPI became dominant in the field because, when they were smaller companies, they were *successful* smaller companies. Most small wargame publishers are *not* successful—and most do not survive their first few years of existence. There are three main reasons for this high attrition rate. The biggest problem is the inability to maintain a sufficient *cash-flow* to pay the bills. This is usually a result of either not selling enough games at a realistic price and/or spending too much money on items that are not really needed. The second biggest problem (which will usually cause the first one, if nothing else does) is the *inability to get, and keep, a sufficient number of customers.* All the careful business planning will do no good if you are unable to obtain a sufficient (planned or otherwise) number of customers. The third problem is an inability to sustain the effort. Many new publishers can go like hell for up to a year on guts and visions of glory. After that exhaustion and extinction set in.

Currently there are two types of smaller publishers, the professionals and the successful amateurs. At the moment, the best example of the professional smaller publisher is Game Designers Workshop (GDW) of Normal, Illinois. To begin with, the people at GDW had the necessary talent in all the aspects of game publishing. They were not initially overly ambitious (which helped the cash flow part), and they proceeded to grow slowly through selling by direct-mail only. Thus they became the third largest wargame publisher in only three years, with sales approaching $200,000 a year.

The only other small publisher with any kind of sustained track record is TSR Hobbies, also from the midwest (Wisconsin). TSR has been around since the late 1960's and has specialized in the fantasy & science fiction division of wargaming. They also deal in miniatures. Indeed, miniatures must be considered at some point, for among the smaller publishers they are heavily integrated with the board type wargames. From a marketing point of view, miniatures gamers (who play no board wargames) comprise some 10% of all wargamers. But in terms of money spent, the miniatures people spend some 30% of the total (some 15% of all wargamers play miniatures). The miniatures figures and equipment are very expensive and time consuming, and this plays no small part in explaining why there are relatively fewer miniatures wargamers and why they account for relatively high expenditure percentage.

There are a number of other small, professional wargame publishers. But at this point the line between the amateur and professional gets a bit blurred, for many of the professional publishers make their actual living in the field by owning and operating a retail store selling wargame materials. Others (like Battleline Publications) might spend time in publishing a small number of games, while earning a living outside the area of wargaming. The part-time publisher (perhaps a more accurate term than "amateur") is the best example of publishing for love, not money, for there usually is very little money involved (a few thousand dollars at most). Given the generally high educational and occupational level of most adult wargamers, it is not unusual to find a number of rather high quality games being published by "private individuals". Since most of these publishers have already established themselves in other careers, few of them get into publishing full time. However, most of the surviving professional publishers were founded by younger gamers who had not yet established themselves in any other line of business. Such was the case with SPI and GDW.

THE PROBLEMS OF PUBLICATION

There are a number of interesting reasons why there have not been many more professional wargame publishers. The field is a highly technical one, requiring considerable knowledge of military history, and history in general, as well as expertise in many aspects of publishing not usually required of a book or magazine publisher. Worse yet, the potential market is quite small, perhaps 250,000 possible wargamers in North America. Reaching these has always been a major problem for the new publisher. Most advertise in the smaller wargame publications (whose circulation rarely exceeds 2–3,000). Ads in *Strategy & Tactics* and *MOVES* magazines are more effective, but also much more expensive (because of the much larger paid circulation, 34,000 and 9,000 respectively). Avalon Hill's magazine, *The General,* has a paid circulation of some 7 to 10,000 (estimated) but has never accepted advertising. In other words, there is not a hell of a lot of money to be made in publishing wargames at this time. And for this reason no major game publisher (e.g., Parker Bros.) has ever made a serious attempt to get into it. This gives the smaller publishers a measure of protection from overwhelming competition, but they are, in turn, left with a woefully small market.

This is not to say that it is impossible to make a profit in publishing wargames. Avalon Hill, after a number of very lean years, posted a record 13% profit (on sales of $1.7 million) in the year ended 4/30/76. During that same period AH accounted for some 20% of their parent corporation's sales while producing some 63% of the total profit. SPI produced a 6% profit on the same level of sales. Like most of the other game publishers, SPI's policy is to produce the most game at the lowest price. This attitude tends to keep prices down, especially in direct mail sales, where there is more competition.

The retail market is dominated by Avalon Hill: they account for some 65 percent of the wargames sold over the counter. The direct mail market, however, is not totally dominated by SPI, even though they account for some 70 percent of direct mail sales. All of the smaller publishers, as well as Avalon Hill, participate in the direct mail market. All of the wargame publications (with the exception of Avalon Hill's *General*) review and advertise other publisher's games. This makes it possible for any amateur publisher to produce and sell up to 1,000 copies of a game. They may not make much money doing this (they often lose quite a bit, plus much time and effort), but such individuals publish for love, not money.

Most of the smaller game publishers do not have highly organized financial records (SPI was, in its early years, probably one of the worst in this regard, growing from $40,000 to $1,000,000 annual volume in four years without a proper set of books). In addition, the smaller publishers tend to pay little or no salaries to their staffs when things get tight. This attitude is not as irresponsible as it appears. Publishing—book publishing in particular, to which wargame publishing is most closely related—has a tradition of placing the quality of the product before the size of the profits. Thus a wargame publisher tends to spend as much, if not more, than is taken in. As an alternative to going out of business from chronic overspending, many publishers resort to various methods of raising investment capital.

There are three main sources for this money-to-grow-with. First, and most obviously, there is direct investment by the publisher or some outside source. Since the prospects of a reasonable return on this investment are slim, the probability of some outside wargame angel coming forth are remote. Similarly, the publisher's own resources are usually limited, although it is usually the publisher's investment which is taking the operation as far as it can. A second very common source of operating funds is the direct mail "float". This term refers to the money that is taken in from direct-mail orders. As many gamers know from experience, seeing an ad for a game does not mean that the game is readily available. As often happens, the mere mention of a new game will bring in orders before the game is ready to ship. Since money is coming in and nothing (at the moment) is going out, this additional money, or "float", is available for other uses. A third source of funds is actually a variation on the second. This is the subscription "float". Subscriptions have a built-in float. You pay the full amount up front and then receive the product over a fixed period of time. A variation on this is the lifetime subscription, which is the ultimate use of a float as a long term investment.

Often these various investments are legitimately needed in order to get the publisher started, buying equipment, building up inventory, buying ads to obtain regular customers, etc. SPI used all three methods, especially the last two, to get started. Most small publishers use the float to get going and, unfortunately, to keep going beyond where they should have stopped. With this last attitude, it's no wonder that publishing is so unprofitable and, as a consequence, so often unsuccessful.

As more publishers survive longer, there is a tendency to pay more attention to profitability and long-term survival. The two usually go together. As this happens, it results in changes in publishing practices which are both good, and bad, for the gamer. On the good side, more games get published. A profitable publisher has an incentive, and the means, to continue publishing while an unprofitable publisher does not. The quality of the games also has a tendency to increase. Anyone who has been buying games for more than five years will readily appreciate this point. In addition, the approach of the publishers tends to become more professional and efficient in general. Now for the *bad* news.

Human nature being what it is, folks get into ruts if they do one thing for any length of time. Unless the publishers can keep their creative people in a "constant state of revolution" (to paraphrase Chairman Mao) their publications will tend more and more towards the bland. Some say this has already happened to both Avalon Hill and SPI. Up until the early 1970's, AH certainly did give many indications of being in a rut. They published only one or two new games a year, games that were, by present day standards, often shabby and uninteresting. Since then, spurred on by other publisher's success in producing dozens of games a year, Avalon Hill has perhaps gotten *out* of a rut.

SPI is quite another matter. Publishing as many as forty new games a year, the volume offended many gamers, even though they were getting games on favorite subjects they thought they would never see in print. While many found what they wanted among SPI's games, they also found much that they didn't like. This was understandable, since SPI showed that the same subject could be covered on many different scales and/or approaches: e.g., *Bastogne/Ardennes Offensive/Wacht Am Rhein,* all covering the Battle of the Bulge; *Cemetery Hill/Terrible Swift Sword,* both on Gettysburg; *Red Star-White Star/Firefight/Mech War '77/Tank,* all on modern armored warfare; and *Napoleon At Waterloo/La Belle Alliance/Wellington's Victory*—all on the battle of Waterloo. Some gamers were irked by this, being disappointed when a subject they preferred in one scale was published in another.

The "mistake" that SPI made, which caused it to draw criticism, was in *raising* the standards of game design and production and consequently raising the *expectations* of the gaming public. Even when SPI designed games for other publishers, its groundbreaking approach simply produced another "hard act to follow". For example, the SPI team designed, and produced all the artwork for Avalon Hill's *PanzerBlitz* (AH's all-time bestseller). And then when SPI produced later tactical armored games that improved upon the system in *PanzerBlitz,* some gamers complained that they weren't as good as "AH's design" even though the same people had done both!

When SPI produced a weak design (whether by truly fouling it up or by reaching too far experimentally) it also drew an inordinate amount of fire from its public. Conversely, some complained (and still do) that SPI mass-produces game designs, cookie-cutter style. The decision to produce a number of games using common or related systems truly sprung from the notion that so many situations *share* so many characteristics that to design totally different systems would be gratuitous. For example, the division-level games on the East Front (*Kursk, Turning Point, Destruction of AGC,* et al.) used essentially the same system for this reason. The QuadriGame series was an attempt to produce related games using easy-to-learn, common systems to widen the audience for the hobby and to provide gamers with low-cost, fast-playing games. SPI has tried to serve the interests of its customers from a broadly-based position. Of course, there are always alternative approaches. A sizeable number of gamers prefer novelty and diversity in *every* game design they purchase, even if this sometimes means a sacrifice in playability and quality (not to mention realism and historicity).

Such diversity of design approach was difficult to achieve at SPI for many years because most of the games were being designed by one person: Jim Dunnigan. Of some 150 games SPI has published from late '69 through early '77, about half were Jim Dunnigan designs. Most gamers are vaguely aware of this, but it is doubtful that many realize the full meaning of so many games being designed by one person. For one thing, whole classes of games (e.g. tactical armored games) were literally monopolized by this one designer. The point to be made here is that Jim Dunnigan is no superman when it comes to game design. No one else has ever designed so many complex published games, particularly wargames, by far the most complex type of manual game available, in so short a period of time. SPI has not planned for that state of affairs to go on forever, and in the past few years the proportion of SPI's new games done by Dunnigan has steadily declined. (Some say this is why the quality has steadily gone up!) The "Plan" at SPI envisions a long-term training program for game R&D people that will continually upgrade the quality and quantity of games. This approach also accounts for many of the strange twists and turns SPI's game design has taken over the years. You find out if something works by doing it.

If you want to find something good, or bad, about games design, chances are you'll find it all at SPI. To be rather blunt about it, the only way SPI avoids falling into a very large rut is by encouraging turnover in its own R&D staff. This is done by holding everyone's efforts up to critical scrutiny via the feedback ratings and budget/schedule performance. Those that can't "cut the mustard" are encouraged to do better. An in-house designer usually either flourishes or wilts under that type of pressure. Eventually, however, the long term, "battle hardened", professional R&D people will make up (more and more) a majority of the R&D staff. At that point, part of their job is to produce substantially better games than the hundreds of quite competent amateur designers. Whether or not this is possible is currently an open question. Historical simulations are now going through a stage which book publishers only passed through fleetingly over a hundred years ago. It's anyone's guess as to exactly what will happen. While many gamers are concerned about this potential problem, the publishers are trying to have it both ways by cultivating outside designers while still using in-house staff. SPI, by actually *hiring* outside designers, is thus engaging in a variation of the same solution.

INCOME AND EXPENSES ON THE GAME *WORLD WAR 3*

Month	Date	Inc	Exp	Status	Unit Sales	R&D Pay	Over-head	MFG	Fulfill	Mar-keting	Margin
1	6/74	—	1836	–1836	—	698	1138	—	—	—	—
2	7/74	31	711	–2516	5	267	435	—	9	—	—
3	8/74	52	1149	–3613	8	432	704	—	13	—	
4	9/74	37	688	–4264	6	257	419	—	12	—	—
5	10/74	121	1602	–5745	19	597	973	—	32	—	—
6	11/74	704	1342	–6383	94	448	730	—	164	—	—
7	12/74	4762	4104	–5725	684	1191	1941	—	972	—	—
8	1/75	7898	5052	–2879	1145	10	16	—	5026	—	—
9	2/75	2792	3403	–3490	1297	52	85	1442	1824	—	—
10	3/75	3854	3476	–4812	929	—	—	1790	1686	1700	—
11	4/75	2361	2137	–4588	524	—	—	880	1257	—	—
12	5/75	4095	1047	–1540	469	—	—	—	1047	—	—
1-12	6/74-5/75	26707	28247	–1540	5178	3952	6441	4112	12042	1700	–5.8%
13-24	6/75-5/76	20184	17996	2998	6588	—	—	5639	12107	250	14.9%
25-36	6/76-5/77	17250	14525	2725	5616	—	—	4500	9775	250	14.4%
1-36	6/74-5/77	64141	60768	3373	17382	3952	6441	14251	33924	2200	5.3%
As % of EXP		—	**100%**	—	—	**6.5**	**10.6**	**23.5**	**55.8**	**3.6**	—
As % of INC		**100%**	—	—	—	**6.2**	**10.0**	**22.2**	**52.9**	**3.4**	**5.3%**

Month: the month number from the time we started working on the game. **Date:** the actual month. **Income:** the cash actually received for the game. This does not exactly correspond to game sales as we are not paid for retail and foreign sales until 60-90 days after the game is sold. **Exp(ense):** the cash actually paid out that month for work on the game. **Status:** taking the cumulative expenses and subtracting them from the cumulative income for the game. In other words, how much we are ahead or "in the hole." **Unit Sales:** how many games have been bought that month, includes direct mail, retail and foreign sales. **R&D Pay:** the payroll for R&D people (including the art department) working on the game. **Overhead:** at SPI overhead (all non-payroll, non-manufacturing expenses, including rent, utilities, insurance, legal and accounting fees, equipment purchases, payroll taxes and benefits, etc) is allocated according to payroll. Each dollar of payroll gets a proportionate share of overhead expense attached to it. Easily overlooked (and often the cause of death for many young companies), it is a considerable expense. **Manufacturing:** the actual cost of manufacturing the game, the plastic box is included in *fulfillment* as part of packaging and shipping. **Fulfill(ment):** the cost (mostly labor, but including the necessary materials) for processing and shipping the order. This cost is highest in direct mail but also applies, to a lesser extent, to retail and foreign sales. **Margin:** the percent of difference of sales over expenses (if it's the other way around you're in big trouble). **Marketing:** the yearly cost, in dollars, for advertising and promotion. Primarily the cost of a full page ad in *S&T* plus regular listing in the SPI brochure. The last two lines of the chart show each expense category as a percentage of total expense (Exp) and total income (Inc).

THE COST OF GAME PUBLISHING

The game *World War 3* was published in December, 1974. Actually, manufacturing of the game was finished (and copies available) on 17 January 1975. As the accompanying table shows, at the time SPI shipped the first copy of *World War 3* they were in the hole for some $5000. The table also shows that things eventually got better.

SPI first began putting money into the production of *WW3* in June 1974. At about the same time word got out that they were doing the game, and a few advance orders began to trickle in. For the first seven months the major expense was R&D (Research and Development, including artwork) and fulfillment (taking orders in and processing them). On the eighth month the nearly 2,000 backed-up mail orders were shipped, which is why the fulfillment costs were so high in January 1975. Since SPI has 30 days to pay their printing bills (and most of their other manufacturing bills) they do not incur any manufacturing expense until February.

The initial run of a game is usually 5,000 copies. Starting in February, most of the games sold are in retail or foreign outlets. On these sales SPI usually does not see any money until at least 90 days *after* the games are shipped and sold. After the first year most of the games sold are in retail outlets. But, because SPI usually paid off their R&D expense in the first year, the lower amount of money received from retail sales does not hurt them.

It is estimated that a publisher will make some 50¢ on each copy sold at retail, $1.60 on each copy sold through the mail or $1.45 for each copy sold to foreign outlets. The pricing policies in retail can be harmful, mainly because you get so little back for your game. When you sell retail you can expect to get back (after discounts and expenses) no more than 30–40% of the list price of your game. For a game like *World War 3* (list price $8.00) this comes to about $2.50. Selling the game through the mail brings in about $6.60, less about $3.00 to process and ship the order, leaving some $3.60. (Selling to the foreign market brings in some $2.45). Once the R&D expense *has* been paid off, you can survive with retail sales. However, it is the large direct mail sales of a game which allows SPI to survive with (relatively) low sales volume per title. A publisher that depends more on retail sales needs larger sales to pay off the R&D bill. Avalon Hill is an example of this. A company that relies more on direct mail can get along with even lower sales per game. Game Designers Workshop is an example of this approach.

The publisher must be effective in controlling costs. If your costs exceed your income you have to make up the difference somehow. If you can't, you're no longer publishing games. *World War 3* was one of over a hundred games SPI has published, most of which have made money. A few have actually *lost* money and some barely broke even. The promotion for these games was mainly centered in SPI's own *Strategy & Tactics* Magazine. Until recently, the magazine was a money *losing* operation, and only the money SPI made on the games kept the whole operation going.

There are four major expenses in publishing a game. First, there is the cost of designing and developing the game. Then comes the cost of manufacturing the game. Third, there is the cost of taking and filling orders. And then there is a fourth, and sometimes overlooked, the cost of marketing (i.e., advertising and promotion). To get a new game started SPI runs a full page ad in *S&T* and, as an internal accounting transaction, bills it to the game as an expense. The game pays the full page rate of some $1500. In addition the game must pay a couple of hundred dollars a year for being in our brochures, etc. The ability to provide an efficient marketing medium for SPI games is the main *economic* rationale for publishing *S&T*.

You don't really get much in the way of bulk when you buy a game: a lot of cardboard and not much printed matter, especially when compared to a book. Actually, the book is cheaper to produce. A nine dollar book costs about $1.50 to manufacture. A nine dollar (SPI) game costs about $1.645 This breaks down to 26.5¢ for a 23x35 inch, three-color map (an extra color costs 2.5¢ more per map); 24¢ for a twenty-four page rules booklet (16 pages goes for 15¢, 32 pages for 29¢, etc.); 28¢ for 400 three-color, back printed counters. (One hundred, three color, back printed counters would cost 7.5¢); 72¢ for the plastic box, with the cover sheet costing some 12¢ and the die and band another 2¢. A "bookcase" type game would cost about the same for the packaging. The additional expense would be in the mounted board. To put a 23x35 inch SPI map on stiff board costs about one *dollar* more. All of the above costs are as of late 1976 and are for quantities of at least 5,000 (the same basic quantities normally used in hardcover book manufacturing).

The usual rule of thumb in pricing games is that the retail price should be *six times* the manufacturing cost. As far as is known, Avalon Hill still follows this practice (primarily to survive the heavy discounts of retailing). SPI uses a 4 or 5 to 1 ratio (based on the average price on direct-mail games, after subscriber discounts). Many of the smaller, direct-mail-only publishers use a 3 or 4 to 1 ratio, partly because of their higher manufacturing costs, which are a result of inexperience and/or low print runs.

In addition to direct production costs, there are also the costs of actually supervising the manufacturing. The manufacturers (printers, die-cutters and other suppliers) are only human. If you don't (or can't) tell them *exactly* what you want you are going to have problems. And these problems are expensive. Another unavoidable "manufacturing" expense is the assembly of all the game parts into complete games. Whoever does it, it's going to cost something, say about 10¢ a game. Finally, there is warehousing (storing the games somewhere until sold). That can easily be another few cents a game, plus a few cents for transporting all this paper hither and yon. Nothing is simple in this life, and the manufacturing costs of a game prove this point very well.

As can be seen from the table, the single largest expense is "Fulfillment". This is the function which gets the game from the publisher to the gamer. Fulfillment is expensive, as many *former* game publishers learned too late. If gamers do not get their games they will not buy any more. If the publisher doesn't take in enough cash to keep the operation going, the operation grinds to a halt. The act of opening the gamers' envelope, counting the money, preparing a shipping label, wrapping up the game, putting postage on it and getting it out the door *seems* simple enough. But this seeming simplicity is shot through with potential hazards. To paraphrase Murphy's Law, anything that can go wrong with a game order, will go wrong, and at the worst possible moment in a manner guaranteed to cost you the most time, money and sweat. Furthermore, it is also the easiest area to ignore (who wants to be working in the mail room when there are games to be designed!) As a game publishing operation gets bigger the fulfillment costs drop drastically, mainly because your fixed costs for space and equipment are spread over a greater number of transactions. In addition, if you're on the ball, bigger can be better as far as being able to handle a larger number of orders more cheaply on a per-order basis. However, this applies only up to a point. Keep in mind that fulfillment also includes the cost of the shipping materials.

The high cost of fulfillment (over 50% of total expense) is largely attributable to packaging, including the shipping container, which consumes some 45% of the fulfillment expense. Postage takes up another 35% while labor consumes most of the rest. Computer and other machine charges take up only a few percent. So be careful about what happens in the "back room".

SPI's manufacturing costs are a little on the high side according to the traditional game publishing rule-of-thumb, which holds that manufacturing cost should be one sixth, or 16.6%, of the retail price. SPI's manufacturing cost (including the cost of the box) is some 20% of the retail price. But this is because SPI sells at least 25% of a game through direct-mail, which could enable a company to survive on even a 33% manufacturing cost because of the higher profit margin. Keep in mind that when you buy a game in a store the game publisher is getting *less than half* of the money you are paying for that game. The rest goes to the "middlemen", the store owner and, quite often, the distributor who takes his orders and delivers the games. When you buy a game through the mail there is no middleman, except for the Post Office or UPS. The publisher must pay someone to process your order as well as rent and overhead on the place where the order is processed. But this is almost always less of a burden than the wholesale discounts given to the stores.

Why do so many of the games continue to be bought in the stores? Better yet, why do the publishers even *bother* with the stores at all? Mainly because the stores can move the goods. How do they do this? Mainly by being where the people are. That's why the publisher has to pay the store owners' rent by selling the games to them at less than half price. The stores are laid out and designed to get people to buy, which is why you rarely get out of one without buying something. The store owners are no dummies. It costs them a lot of money to get set up and if they don't get sales they go out of business. On the other hand, from half to two-thirds of the wargames bought in stores are bought by people who are not wargamers. While this means nothing to the store owners nor publishers it does mean something to the gamers, because a game designed to sell well in a store will not always be designed to serve the gamers' interests. Often a title will be chosen first, and then a game will be designed around that title. The result is not always favorable for the gamer. Sadly many of these games bought by non-gamers are never really used.

Let's face it, wargames are very complex. The best selling wargame of all time, *PanzerBlitz,* is quite complex. It also has stunning package graphics (designed by SPI's Simonsen) and a pretty zippy title. The game has sold well to gamers and non-gamers alike. Unfortunately, most of the sales were probably to *non*-gamers who, after trying to play *PanzerBlitz* unaided, most likely decided to *remain* non-gamers. Another problem with the marketing of games is that approximately half the population will not buy anything by mail, with the exception of magazine subscriptions. Thus the stores have half the population all to themselves. And then there is the convenience of the stores, even for those who do buy through the mail. Gamers, even in the face of the lower prices direct mail sellers can offer, still like to buy in a store. The point of all this is that in marketing, as in so many other things, logic is not the governing force. Sometimes it's just being there that counts.

From a gamer's point of view, the marketing methods of publishers have a great effect on how well the gamer is served. And the wargamer is part of a relatively small group (100–150,000 active gamers, at most). But initially, when wargames were first introduced in the late '50's it was through the standard toy distribution system. Toys are a mass market. Toys and games are judged a success if their sales number in the hundreds of thousands of units. A game, being a little more specialized, could do all right with sales of less than 100,000, especially if it continued to sell (unlike most games and toys) beyond a year or so. This large volume also created larger discounts (you might only get 30–40% back from a store), heavy competition and considerable risk for the smaller publishers (Avalon Hill went broke in this type of market during 1964). Operating in this market produced the "one or two games a year" system Avalon Hill maintained during their monopoly years (especially 1964–73). This was hard on the gamers in two ways: there were not many games, and those being published were not always outstanding (some were out-and-out bombs).

Thus selling in stores is a two-edged sword. On the one hand the game will receive wide distribution. On the other hand the gamer, except in a few discount houses, will have to pay full price; the games will be published more on the strength of title appeal than anything else; there will be a tendency to avoid "saturating the market" with too many new games; only a narrow range of subjects will be considered in order to insure wide appeal; and, finally, the publishers will be selling initially to the retailers, who know little about the games and mainly want an eye-catching box. (That's all you have to present at trade shows; they don't really care what's inside).

The other marketing approach is to sell the games like books, through book distribution channels. The people who buy games tend to buy many hard cover books each year and are more prone to visit book outlets than toy stores. In addition, the normal "trade book" (hardcover, selling for $8–15+) only sells 5–7,000 copies. Thus if the game publisher can gear sales and publications costs to fit a smaller manufacturing run, many beneficial things will occur.

First of all, more games can be published. There is less risk in each of the more numerous games put out. Secondly, there can be much less reliance on "popular" subjects. Thirdly, producing a larger number of games allows for the use of professional staff, other than part time or free-lance people. This has the advantage of enabling long range R&D projects, new game systems, and advancements in the state of the art in general. Finally, the efficient use of smaller publishing runs allows for less commercial pressures in general. More attention can be paid to designing and publishing a quality *game* instead of all the hustle involved in selling a *product* to the retailer. The small-run publisher is a lot closer to the gamer and, best of all, the small-run system works. Indeed, it has increased the number of wargames sold yearly by seven times since it was introduced in 1969.

Even Avalon Hill, which pioneered commercial wargame publication in the late 1950's and had always sold their games through the toy distribution network, finally changed direction in 1974. Two years later AH was able to break a three year stagnation of sales near zero growth with a 34% increase and a 13% profit margin. They produced more games in 1975–76 than they had in the previous ten years combined. Avalon Hill also went deeply into direct mail, which they had previously played down for fear of offending their retail distributors. They also turned a vice into a virtue by using their royalty system to buy up already published games of the smaller game companies and then republish them as Avalon Hill games. This was a variation on what Avalon Hill had been forced to do most of the time since 1966, i.e., get most of their games from outside designers. Nearly half of their game releases published between 1967 and 1973 came from SPI.

The vast proliferation of published games after 1969 also produced a proliferation of smaller game companies, each doing 1–5% as much business as SPI or AH. When Avalon Hill decided to get on the band wagon and publish more games they quickly realized that it could not be done with their nearly non-existant R&D department. So they simultaneously began an R&D department who's main function was to take the smaller publishers games and "clean them up", repackage them and publish them as Avalon Hill games. Within a few years they were able to publish as many as 6–8+ games a year. The royalties paid were not that great, usually some 1 percent of the retail price of the games, in comparison with 10 percent paid to book authors and the 2½ percent that the SPI crew got for most of their Avalon Hill designs. Still, even at 1 percent, Avalon Hill was paying out a substantial amount for game designs. Take, for example, a nine dollar game that sold 50,000 units over its published life. That's $4500. Add to that the R&D labor and overhead (say, at least a thousand dollars) and the same for art work and you have a six to seven thousand dollar cost, which is about what it costs SPI to do the same job in-house. The royalty method has other advantages. If the game doesn't sell well, you pay less royalty. Moreover, you don't have to pay out the royalty part until *after* the game is sold, and you aren't stuck with a large (and expensive) R&D staff and overhead cost if business should turn bad. It is difficult to simply fire people, and if you do, they're going to be hard to replace. You also have to train them in-house, and that's an expensive process. On the other hand, the in-house work done by Avalon Hill can get out of hand (expense wise) and drive the ultimate cost over the equivalent cost for a staff design. When carefully managed, both systems work.

The smaller game companies, believe it or not, often use both systems (either/or, or mixed). Anyone getting royalties from a smaller publisher isn't getting much (a few hundred dollars at most) but then the regular staff of the smaller companies don't get paid much either and, in many cases, nothing at all. Face it, most aspiring game designers would gladly take nothing if they could just get their game published in a professional style. As we have seen, the R&D expense of publishing is rather minor. This is rather dismaying to the designer, but it is an economic fact of life one must live with. Games are still being published in defiance of all the economic facts of life. This is how game publishing had its beginnings. As long as the urge to publish is stronger than the urge to get rich, there will be games published. As the old saying goes, you can make money or you can make sense.

How Many People Play the Games?

The number of actual wargamers is difficult to estimate. We know how many are sold, but unless we know how many each gamer buys it is difficult to compute the number of gamers. Using the data SPI has compiled over the past five years it appears that there were no more than 20,000 active gamers up until 1970. This figure is based on first year sales of new games by Avalon Hill. These never exceeded 26,000, and it appears that less than half of these sales (almost all through retail outlets) were impulse sales to non-gamers. The number of gamers seems to have been growing slowly, mainly because so many people left gaming for any number of reasons. There were, however, a large number of semi-active gamers: people who had once been gamers, were no longer actively involved, but could be easily re-activated under the proper conditions. The proper conditions appear to have been more games and more advertising. SPI began advertising on a large scale in late 1971 and by late 1972 had acquired over 20,000 subscribers. Many of these commented that they had been out of gaming for lack of interesting games.

It is known that the average *S&T* subscriber buys some seven games a year. That, plus the games delivered in *S&T* subscriptions, comes to some 400,000 games a year. Assuming that the remaining 100,000 games are bought by somewhat less devoted gamers gives us another 20,000 people. That's some 50,000 very active gamers. We know that more that 75,000 people read each issue of *S&T*. We know also of large numbers of gamers who game but do not buy games. This brings the number of gamers up to the 100,000–125,000 range. It's a mixed lot. On one extreme we have some 5,000 "collectors" who buy just about every game available. Somewhat larger than this group are the "grognards" (long term, veteran, gamers); we estimate there are 10,000–15,000 of these folks. Then there are another 30,000 or so regular gamers, and beyond this we have some 50,000–80,000 gamers who may, or may not, play with any regularity.

The important question that remains unanswered is why people get into, or out of, the games. There aren't many people in this hobby, there are very few "casual" gamers. Perhaps the reason is that it requires too much concentration and energy simply to understand the rules and use the games. All we are certain of is that gamers are a relatively small group, supplied by relatively small publishers, all hoping that smaller *is* better.

WHAT DOES THE FUTURE HOLD?

Aside from the fairly safe prediction of many more games, the future is clouded by the shifting mist of technology and sociology. It may be possible, within the next ten to twenty years, through the magic of computers, with an assist from the television industry for "wargames" to be playable on computerized TV sets, or variations on that basic theme. This just might produce a significant setback for the "manual" wargames publishers. At least one of the major wargame publishers (SPI) is already working on this aspect of wargames, more out of curiosity than self-defense; SPI would not likely be able to match RCA in a marketing competition. The above could be the wave of the future, or a mirage. Or a little of both.

And then there is sociology. Consider the following: most gamers (some 92%) are under the age of 34. This has been true for as long as anyone has been keeping track of such things. In the next ten years the enormous post-World War II "baby boom" group will be passing beyond their mid-thirties. In effect, the present pool of potential wargamers will contract. Should this happen, the game publishers would be profoundly affected. They would lose many of their customers and be thus forced to contract their operations. Under these conditions, it would not be unusual for many of the weaker companies to disappear all together. There would be fewer games published and less of everything in general, unless the active age of wargamers can be extended beyond their mid-thirties. No one is really expending a much effort on this problem at the moment. There may not be a great deal anyone can do about it. The potential for a major problem is there.

Aside from the above (potential) major problems, there are a number of other, more interesting, trends. First, there has been a tendency to publish more and more games each year, with fewer of each game being sold. This could eventually have the effect of making the game business even less profitable than it already is, except for amateurs. The existing publishers will have to adapt to these new conditions.

In 1969 there were only some 100,000 wargames sold. Over half of these were bought on impulse by non-wargamers. Thus, eight years ago, only some 50,000 wargames (at most) were *deliberately* bought. In 1976 nearly 800,000 wargames were sold. And nearly 500,000 of them were bought deliberately. Two years ago SPI's market research folks saw the beginning of a slackening off of sales, and this trend has since become more pronounced. This leveling-off of overall sales, coupled with the publication of more titles, has a number of interesting implications. For one thing, it becomes easier for smaller publishers to get started. Gamers now demand a wider range of games, and more gamers are what we call "collectors". These people no longer, with a few wealthy exceptions, buy *every* game published (nearly a hundred a year in 1975–77). But there are at least a thousand gamers that will buy just about any individual game, no matter how obscure. Using the smaller gamer magazines initially, and eventually *S&T,* many small publishers have gotten their start. In addition, nearly all gamer publications, including *S&T* (with over 75,000 readers), review nearly all wargames that come to their attention.

The smaller publishers rarely make the transition to anything larger. But they, as well as the few established small publishers, have the effect of keeping the larger publishers (mainly AH and SPI) "honest". Just as SPI grew because of what Avalon Hill didn't do, the same could happen to SPI (and once more to AH). The publisher that listens more closely and accurately to the gamers will get the most gamer attention. The openness now prevalent among wargame publishers cannot be easily reversed. Moreover, the average over-21 gamer is college-educated, with half of this group having some graduate school. Given this atmosphere, the people involved in historical gaming—both publisher and gamer—are capable of any number of interesting activities.

SECTION VII:

Wargame Directory and a Suggested Library of Games

by Richard H. Berg

The list of wargames is courtesy of Jeff Pimper and Bob Zabik, who have compiled and now maintain a computerized listing of all available wargames. Information about this service may be obtained from Jeff Pimper, 347 Michell St., Livermore, CA 94550.

A BASIC WARGAME LIBRARY

It is not too easy to say exactly why people play wargames except to say that the pleasures obtained thereby are wholly subjective. Most gamers will readily agree that there can be a distinction made between a game that is "good" and a game that is "fun to play;" i.e., your favorite games may not necessarily be those games that you consider best in terms of design, ingenuity, interest or whatever.

An example of this dual standard would be that old classic ("classic" meaning any game produced by Avalon Hill prior to 1965 that did not self-destruct after purchase), *Stalingrad,* the first Russian Front game and an inevitable favorite at player get-togethers. *Stalingrad* is a great game; but it is not a good wargame, and it ranks even lower as an historical simulation. Its total blindness to any semblance of history—it has all the historical accuracy of "Machiste versus The Son of El Cid"—and its dreadful game map—which, for all its terrain, may as well be the Mojave Desert—label it as an anachronism in the present state of the art.

On the other hand, *El Alamein,* SPI's little-lamented bomb of several years ago, has a great deal of historical accuracy. Unfortunately, it is not much of a game, garnering about as much playing enjoyment as carting away bodies during the Black Death. Another example of this sort of player futility can be seen in Rand's original version of *Lee vs. Meade.* There, a somewhat interesting system was marred by a game that was so balanced that neither player ever did anything for fear of blowing the whole battle on one die-roll.

What, then, is a good wargame? Unfortunately, that is a little like asking what makes a good book or a great opera. Disagreement abounds as to the particulars, but everyone will say they know one when they see one. One could say that a good wargame combines, in varying degrees, what could be termed the "Candide Approach," or the best of all possible worlds. The good game allows the "player-gamer" to have a few hours of relaxing fun; it provides the historian with insight into the particular subject; and it allows the "assassin," the killer-gamer, to vent his spleen on his opponent by revealing the Master Plan that will not only destroy his opposite number, but History as well.

So far, there are few (if any) games that could satisfy all three of these groups, so we must lower our sights a bit and look to the individual merits of each of the over 250 games that are readily available. Is there something about the game that raises it above the others? Is there something about the game that makes it more desirable to have or more representative of its genre? Is it the definitive game on the subject, or is it just plain, good fun?

The following list is a compilation of games that I personally feel any gamer with more than a passing interest in historical simulations should have. There are games that are not only playable, but most typically exemplify the state of the art today.

The Basic Wargame Library is organized in groups according to the level of the simulation: Tactical, Operational, Strategic, and Multi-Player/Diplomatic. Within each group each game is listed in approximate order of complexity. In addition, the name of each game is followed by the designer and publisher. Addresses for the publishers, for those who wish to obtain more information, can be found at the end of this Section. It should be pointed out that the exclusion of a game from this list does not mean per force that it is a bad game—it simply does not represent the "Best of Wargaming."

Tactical Level Games

Musket and Pike (John Young/SPI) This is a remarkably clean tactical game of the 16th and 17th Centuries that conveys a solid feel for the period. The scenarios are all interesting and they are rated for complexity and balance, something that is rarely, if ever, done.

Soldiers (Dave Isby/SPI) This is the only game from WWI on the list, and it is a remarkable achievement in tac games. A surprisingly freewheeling game that provides solid, historical insight into the first months of the Great War. The scenarios are varied and colorful.

Manassas (Tom Eller/available from GDW) A minor gem from an "amateur" designer. A SiMove system that concentrates on formation and operational-level tactics. A tense contest of wits, backed by solid historical research.

Caesar (Robert Bradley/reissued by Avalon Hill) This first professional publication of the legendary Alesia game has a lot going for it. An exciting double envelopment situation leads to much tension and swift maneuver. A big game for 2, 3, or even 4 players.

Siege of Jerusalem, 70 A.D. (Fred Schacter/Historical Perspectives) An unusual subject handled in an interesting fashion. An "amateur" game, yet a professional feel. Excellent re-creation of siege tactics that maintains a feeling of movement that is unusual for the subject. Quite colorful and very tense.

Frigate and *Wooden Ships and Iron Men* (Jim Dunnigan/SPI; Craig Taylor/Battleline and AH) This is a double entry, for the two games (which can be played together, believe it or not) represent the best naval games on the market. *Frigate* concentrates on maneuver, while *Wooden Ships* leans heavily on the romance of ship-to-ship action. Both games have plenty of swash in their buckles.

Panzer '44 and *Mech War '77* (Jim Dunnigan/SPI) Another double entry, this time because the games are so similar. Probably the best mechanized warfare tac games on the market; a quasiSiMove system limits paperwork, and the scenarios and mapwork are exceptionally realistic.

PanzerBlitz (Jim Dunnigan/AH) The grandaddy of all tactical armored games, one of the most popular games of all time. The mystique of the tank is brought into your house in a glorious physical package designed by Redmond Simonsen. A bit outdated, but never out of style.

Air Force (S. Craig Taylor/Battleline, available through SPI) The best air game on the market, it recreates WWII aerial combat on virtually every level. Bombing missions, dog-fights, V-2 rockets, and so on. The rules are complete yet not overwhelming, and the system is relatively clean considering the information it handles. Well done.

La Bataille de la Moskowa (Lawrence Groves/Martial Enterprises, available through GDW) The first large-scale Grand Tactical game, and the one that showed that it could be done. A blissful marriage of miniatures and boardgames in a delightfully simple system by a first-time designer. Grand fun.

Sniper! (Jim Dunnigan/SPI) The nitty-gritty tactical game and one of the best fun items ever. A consistent best-seller the idea of reducing each counter to an individual proved to be just short of inspirational. The closest thing to actual enlistment going.

Terrible Swift Sword (Richard Berg/SPI) The ultimate Gettysburg game, and the most successful "big" tactical game to date. Regimental level, it covers different weapons and different leaders in a system that is remarkable for its ease of play. To date, it is the highest rated wargame in the industry.

Wellington's Victory (Frank Davis/SPI) This game does for Napoleonics what *TSS* did for the Civil War. An expert simulation of Napoleonic tactics (at Waterloo), the game uses an exciting new combat system to go with some top research work in unit maneuverability and morale. A must for anyone interested in this period.

Operational Level Games

Borodino (John Young/SPI) A remarkably

good game, and one of the oldest on the list. An excellent way of introducing people to wargaming; the simplicity of the play belies an intricacy of strategic possibilities. Fast-moving and exciting.

Chickamauga (Irad Hardy/SPI) One of the finest one-on-one player games around, this game virtually "made" the Folio concept. Retaining an exceptional degree of historical accuracy, it is truly a *rara avis* in that it is an unusual game that is not decided until the very last turns.

Quebec, 1759 (J. J. Dagliesch/Gamma Two) This is a delightful game that uses an unusual system, resembling—for lack of a better term—dominoes. One of the most beautifully crafted games on the market, it is also a very exciting one. Colorful, clean and very playable, with simple, well-written rules.

Arnhem (Jay Nelson/SPI) Somewhat of a sleeper hidden in the *Westwall* quad, this operational simulation of the ill-fated Allied paradrop in Holland is another minor gem. Like most Folio games, it is fast-moving and tense; but it also retains a remarkable degree of historicity, raising it above the level of its brethren.

Lee Moves North (John Young/SPI) *LMN* is an unusual game in that it places a premium on *avoiding* conflict. It epitomizes the different styles of warfare in the Civil War. With the emphasis on maneuver, the hidden movement, limited intelligence system provides an excellent background for this tough battle of wits.

La Grande Armee (John Young/SPI) This is the best of the Napoleonic games, an insight into the speed and power of Napoleon's Army in its heyday. Sweeping movements and sudden battles are complemented by excellent supply and cavalry rules. Each scenario is a major game in itself. One of the best.

Panzergruppe Guderian (Jim Dunnigan/SPI) SPI's best magazine game in years, this game "made" untried units a standard item for games to come. Using a unique CRT and some excellent rules concerning German tactics, *PGG* recreates operational campaigns in Russian as no other game has done before. A player's delight that retains a high level of historicity.

PanzerArmee Afrika (Jim Dunnigan/SPI) The most unusual and, ultimately, the most interesting of the treatments of the well-covered African Campaign. Intricate supply rules combine with some unique counter-attack provisions and those 60 movement point units to provide a highly playable game with good historic insight.

Crimea (Frank Chadwick/GDW) GDW is known for producing some happily off-the-wall items, and this is one of their best. Colorful and creative, *Crimea* recaptures the essence and futility of one of the most mismanaged campaigns of modern (or semimodern) warfare. The basic system uses operational-level movement interwoven with a tactical combat system, and this is the only game of its kind that I have ever seen pull this off.

Desert Fox (Chuck Lane/Cavalier, possibly out of print) One of the best amateur designs of recent years, *DF* is literally a goldmine of ideas. It also is the most historically accurate of all the African Campaign games, with a meticulous set of supply rules and a top-notch system for refitting units. Despite its length, it is quite interesting to play and deserves a much better fate than it has so far received.

Sinai (Jim Dunnigan/SPI) This was a tough choice, considering the excellent competition (John Hill's *Bar Lev*). However, *Sinai* accomplishes more on its operational level than does *Bar Lev* on its quasi-tactical. It is also a lot of fun to play, and a tremendous challenge to both sides. One of the most popular SPI games.

Frederick the Great (Frank Davis/SPI) This is one of the finest game designs of all time, a masterpiece of fusion between game balance and historical perspective. Virtually every aspect of the game is new and fresh, from the interwoven play sequence down to the rules for surrender and Honors of War. A must for any serious gamer.

Narvik (Frank Chadwick/GDW) Who would have thought that a game on the invasion of Norway could hold any interest? This game, however, is one of the best "invasion" games ever designed, and an insight into the problems of logistics and organization. *Narvik* borrows some from the *DNO* system, but is still its own game, with some highly unusual and innovative rules. And the counters are gorgeous!

Anzio II (Dave Williams; redesigned by Tom Oleson/AH) The revision of *Anzio* could possibly be the best game in the AH line. The inherently interesting situation remains, but the anomalies of the first version are gone. In its place is an incisive game that plays differently every time you tackle it, a recommendation for any game.

Fast Carriers (Jim Dunnigan/SPI) Big and brutal, possibly cumbersome and complex, this is still the finest game extant on air/sea warfare. Covering all three levels of war—strategic, operational, tactical—and integrating them into one huge play sequence, *Fast Carriers* recreates all of the major, and most of the minor problems of conducting this type of war. A solid historical simulation.

Drang Nach Osten/Unentschieden (Frank Chadwick/GDW) If you are going to get one of these big games, you might as well try this. Not as playable as its "sister-in-arms" *War in the East,* it is still a remarkable design achievement, with as much color and historicity as anyone could possibly want. The air game itself is overwhelming. It may not be the best game of all time, but it sure is trying the hardest.

War in Europe (Jim Dunnigan and Irad Hardy/SPI) The biggest and—at $40—the most expensive wargame ever designed and published professionally. Despite its size, it is remarkably playable, and it is played quite often. It encompasses the entire WWII in Europe, and Africa, on a weekly basis, including production and the air war. Absolutely fascinating, and a continual source of fun and information.

Strategic Level Games

The Russian Campaign (Jonathan Edwards/AH) Australia's contribution to the gaming hierarchy, this delightfully produced game is a fine portrayal of the Russo-German War at a strategic level. Despite the scale and the relatively fluid ease of play, *RC* still maintains a good grip on history and presents a good challenge to both players.

Chaco (Marc Miller/GDW) Esoterica reigns supreme in this marvelous little item from GDW. With each turn covering a year, this simulation of South American warfare may seem a bit simplistic at first, but the system is subtle and cunningly contrived and both players are hard-pressed to simply beat their opponent into the ground. One of GDW's earliest and most successful designs.

Stellar Conquest (Designed and distributed by Metagaming Concepts) Far and away the best sci-fi game on the market. The basic game is simple, yet effective, and the advanced game and optional rules are quite intelligent and intriguing. As the game progresses, it becomes more and more complex, as the players increase their capabilities—a system which is highly playable and thoroughly delightful.

Conquistador (Richard Berg/SPI) A most unusual game on a most unusual subject—colonization of the Americas. Not truly a military game, the thrust is towards economic survival and greed. Through the use of actual historical personages and a liberal dose of random events, unique rules and a tremendous feel for the period are developed. An ambitious and well-written game.

The Rise and Decline of the Third Reich (John Prados/AH) All right, the rules are garbled almost beyond comprehension; but the game is so much fun. Some excellent and innovative ideas in terms of CRT's and production usage, have been woven into a highly playable, if somewhat ahistorical, version of WWII.

Multi-Player/Diplomacy Level

Diplomacy (Alan Calhamer/Games Research; available through AH) The grandaddy of all diplomacy-type games, and a cult unto itself, this is one of the most popular games of all times. Virtually devoid of any luck—or randomness—the play depends entirely on the wits and guile of the players. There are whole magazines devoted to this phenomenon!

Kingmaker (Andrew McNeil/AH) The first diplomacy-type game since *Diplomacy* to capture the public's fancy. Now in its third version, the game is a wonderful bit of design serendipity where even the vagaries can be fun. Stunningly beautiful—perhaps the most eye-pleasing game on the market—*Kingmaker* is rapidly gaining enthusiastic backing.

A DIRECTORY OF GAMES

The following constitutes a relatively comprehensive list of all wargames available and some that are *not* available (included for reference purposes). The list covers mostly published games and tends to avoid variants and other alterations to games. This list is not absolutely definitive. Aside from those omitted due to irreducible error, there are about one hundred more very obscure titles that are not listed here—titles of interest to none but the collector of the most arcane.

Each game is listed alphabetically with its publisher and year of publication. Games with numerical-year titles are at the end of the list. An asterisk (*) denotes a game is no longer in print. This does not mean that a copy may not be found; collectors are everywhere who might be willing to part with one. Finding these people can be a problem however.

Following the main listing is a list of some of the game publishers. Inquiries about specific games may be sent to them; as for games whose publishers are not listed, a little detective work may be necessary.

A free introductory game is available from SPI (see list of publishers). A copy of *Strike Force One* will be sent to anyone requesting a copy. Simply write to SPI.

Game Title	Publisher	Year of Publication
ACAV (PanzerBlitz Variant)	Jagdpanther	73
*ACW**	AHIKS	68
African Campaign	Jedko	74
Afrika Korps	Avalon Hill	64
*Afrika Korps II**	Panzerfaust	71
After the Holocaust	SPI	77
Airborne	Jagdpanther	75
Air Force	Battleline	76
*Alamo, The**	Conflict Mag.	74
Albania	Taurus	75
Albuera	Perfidious Albion	75
Alesia (See Caesar)*	Robt. Bradley	70*
*Alexander the Great**	Guidon	71*
Alexander the Great (2d Ed)	Avalon Hill	74
Alexander's Other Battles	Panzerfaust	72
Alien Space	Lou Zocchi	73
American Civil War	SPI	74
American Revolution	SPI	72
Ancient Conquest	Excalibre	75

Antietam	SPI	75
Anvil-Dragoon	Jagdpanther	76
*Anzio**	Avalon Hill	69
Anzio (2d Ed)	Avalon Hill	74
Anzio Beachead	SPI	70
*Arbela**	Gygax	69
Ardennes Offensive	SPI	73
*Ardennes, 1944**	ADA	72
*Armaggedon**	SPI	72
*Armored Assault**	ADA	72
Arms Race	AIWA	76
*Arnhem**	Panzerfaust	70
*Arnhem**	Spartan	71
Arnhem (Westwall Quad*)*	SPI	76
*Assassin**	SDC	72
Atlanta	Guidon	73
Atlantis, 12,500 BC	Excalibre	76
*August, 1914**	Jewett	29
Austerlitz	SPI	73
Avalanche (Salerno)	GDW	76
Banzai Blitz	Spartan	76
*Barbarossa**	Poultron (SPI)	69
Barbarossa (2d Ed)	SPI	71
Bar Lev	Conflict Games	74
Bastogne (Westwall Quad*)*	SPI	76
Bastogne, Battle of the Ardennes	SPI	70
Bataan	Balboa	72
*Battaille de la Moskowa, La**	Martial Enterprises	75
Battaille de la Moskowa, La (2d Ed)	GDW	77
Battle Fleet Mars	SPI	77
Battle for Andromeda	Taurus	75
Battle for Germany	SPI	75
Battle for Hue	SDC	73
Battle for Madrid	Jagdpanther	76
Battle for Midway	GDW	76
Battle of Brandywine	Oldenbourg Grenadiers	76
*Battle of Britain**	Gamescience	68
Battle of Britain (2d Ed)	Lou Zocchi	71
Battle of Britain Campaign	Decalset	?
Battle of Camden	Oldenbourg Grenadiers	76
Battle of Chickamauga	Flying Buffalo	74
Battle of Helm's Deep	Fact & Fantasy	74
Battle of Moscow	SPI	70
Battle of Nations	SPI	75
Battle of Saratoga	Oldenbourg Grenadiers	76
*Battle of Shiloh**	Spartan	71
Battle of Slag Hills	Fact & Fantasy	75
Battle of Trenton/Princeton	Oldenbourg Grenadiers	76
Battle of the Atlantic	Taurus	75
Battle of the Bulge	Avalon Hill	65
*Battle of the Five Armies**	Larry Smith	75
Battle of the Five Armies	TSR	76
Battle of the Hydaspes	Avalon Hill	75
Battle of the Pyramids	Northwind	76
Battlecruiser	Eagle Games	75
Battles for Tobruk	Balboa	75
Battles of Bull Run	SPI	73

Battles of Patton	Research Games	73
Battle Wagon Salvo	Lou Zocchi	75
Bay of Pigs	Jim Bumpas	76
BEF (1940 Variant)*	SDC	73
Belle Alliance, La	SPI	76
Bismarck	Avalon Hill	62
*Blitzkreig**	Avalon Hill	65
Blitzkreig (2d Ed)	Avalon Hill	75
Blitzkreig Module System	SPI	69
Bloody Ridge	SPI	75
Blue & Gray	SPI	75
Blue & Gray II	SPI	76
Borodino	SPI	72
Brandy Station	Rand	76
*Breakout**	Third Millenia	73
Breakout & Pursuit	SPI	72
*Breitenfeld**	Panzerfaust	70
*Breitenfeld**	ADA	72
Breitenfeld	SPI	76
Bulge 1940 (Variant)	Avalon Hill	75
Burma	GDW	76
"CA"	SPI	73
*Caesar (*2d Ed of *Alesia)*	Avalon Hill	76
*Casear's Legions (*2d Ed of *Eagles!)*	Avalon Hill	75
Cambrai 1917	Rand	74
Campaign	Waddington	
Campaign of 1876	Northwind	76
Campaigns of Gen. MacArthur	Research Games	74
Cauldron (North Africa Quad*)*	SPI	76
Cemetery Hill	SPI	75
*Centurion**	SPI	71
Chaco	GDW	73
*Chancellorsville**	Avalon Hill	61
Chancellorsville (2d Ed)	Avalon Hill	74
Chancellorsville Variants	Avalon Hill	76
Chariot	SPI	75
Chattanooga	SPI	76
*Cherbourg**	Panzerfaust	71
Chicago! Chicago!	SPI	70
Chickamauga	SPI	75
Chinese Farm (Modern Battles Quad*)*	SPI	75
Citadel	Fantasy Games	76
*Civil War**	Avalon Hill	61
*Cold War 2007**	ADA	71
Combat Command	SPI	72
Combined Arms	SPI	74
Confrontation	Gamescience	71
Conquerors, The	SPI	77
Conquistador	SPI	76
*Coral Sea**	GDW	74
Coral Sea (2d Ed)	GDW	76
Crete	SPI	69
Crete	Taurus	75
Crimea	GDW	75
Cromwell	SDC	76
Crusader (North Africa Quad*)*	SPI	76
Custer's Last Stand	Battleline	76
*D-Day**	Avalon Hill	61
D-Day (2d Ed)	Avalon Hill	65

*Dark Ages**	SPI	71
*Decline and Fall**	Terence Donnelly	72
Decline and Fall (2d Ed)	Wargames Research	74
*Deployment**	Poultron (SPI)	69
*Desert Fox**	Cavalier	72
*Desert Rats**	Spartan Journal	73
Desert War	SPI	73
Destruction of Army Group Center	SPI	72
Dien Bien Phu	SDC	73
*Diplomacy**	Games Research	61
Diplomacy (2d Ed)*	Games Research	71
Diplomacy (3d Ed)	Avalon Hill	76
Dixie	SPI	76
Drang Nach Osten	GDW	73
Dreadnought	SPI	75
Dungeon	TSR	75
Dungeons and Dragons	TSR	74
*Dunkerque, 1940**	SDC	72
Dunkirk, The Battle for France	Guidon	71
Dynasty	Dynasty Games	70
Eagle Day	Histogames	73
Eagles! (See Caesar's Legions)*	GDW	73
Early Pacific Battles	Bruce Moore	74
EastFront	Control Box	76
East is Red, The	SPI	74
Edgehill, 1642	Excalibre	76
El Alamein	SPI	73
Eldritch Wizadry V & L	TSR	75
Emperor of China	Dynamic Design	72
*Empire I**	Third Millennia	72
Empire of the Petal Throne	TSR	75
English Civil War, The	Ironside	76
Escape from Colditz	Parker	73
Fall of Bataan	Jagdpanther	75
Fall of Rome	SPI	73
Fall of Tobruk	Conflict	75
Fast Carriers	SPI	75
Field Marshall	Jedko	76
Fight for Tobruk	Wellington	76
*Fight in the Skies**	Mike Carr	68
Fight in the Skies (2d Ed)*	Guidon	70
Fight in the Skies (3d Ed)	TSR	75
Firefight	SPI	76
First Indo-China War, The	Third Millennia	73
First World War, The (War in Europe Mod 1)	SPI	77
Flight of the Goeben	SPI	70
Flying Circus	SPI	72
*Flying Fortress**	Poultron (SPI)	69
*Flying Tigers**	Poultron (SPI)	69
*Flying Tigers II**	Third Millennia	73
Formalhaut-II	AWA	75
Fortress Rhodesia	Michael Raymond	76
Foxbat and Phantom	SPI	73
*France, 1940**	SPI	71
France, 1940 (2d Ed)	Avalon Hill	72
Franco-Prussian War	SPI	72
Frederick the Great	SPI	75
Fredericksburg	SPI	76

Freiburg (Thirty Years War)	SPI	76
Friedland 1807	Imperial	75
Frigate	SPI	74
Galactic War	Usborne Pub.	76
Gar-Garouk	Eagle Games	75
Gettysburg (Original Grid)*	Avalon Hill	58
Gettysburg (2d Ed; Hex)*	Avalon Hill	61
Gettysburg (3d Ed; Grid)	Avalon Hill	64
Global War	SPI	75
Golan	SPI	75
*Gorlice-Tarnow**	Jagdpanther	74
Gotterdamerung	Jagdpanther	75
*Graf Spee**	Third Millennia	73
Grande Armee, La	SPI	72
*Granicus**	ADA	72
Grenadier	SPI	71
*Grunt**	SPI	71
*Guadalcanal**	Avalon Hill	66
*Guerre a Outrance**	SDC	72
Guerrilla	Maplay	75
H-Hour	Balboa	76
*Hannibal**	Larry Rusiecki	69
Hannibal (2d Ed)	Histogames	72
Highway to the Reich	SPI	77
Hitler's Last Gamble	Rand	76
Hooker and Lee	SPI	76
*How the West Wasn't Won**	ADA	72
Hull Down	Alnavco	69
Hurtgen Forest (Westwall Quad)	SPI	76
Imperialism	Flying Buffalo	74
Indochina	ADA	72
*Invasion**	ADA	71
*Invasion of Sicily**	Panzerfaust	72
Invasion: America	SPI	76
Invasion: Sicily	Rand	74
Invasion: Sicily (2d Ed)	Gamut of Games	74
Island War Quadrigame	SPI	75
*Italy**	Poultron (SPI)	69
*Iwo Jima**	Panzerfaust	70
Jacksonville, The Beaches of Doom	Jagdpanther	76
Jena-Auerstadt	SPI	75
Jerusalem	SDC	75
*Jutland**	Avalon Hill	67
Jutland (2d ED)	Avalon Hill	74
Kampfpanzer	SPI	73
Kasserine	SPI	76
Kasserine Pass	Conflict	73
Khalkhin-gol	SDC	73
*Kingmaker**	Philomar	74
Kingmaker (2d Ed)*	Philomar	75
Kingmaker (3d ED)	Avalon Hill	76
Knights at War	Usborne	76
*Komet und Meteor**	Jagdpanther	74
*Korea**	Joe Angiolillo	72
*Korea**	SPI	70
Korea (2d ED)	SPI	71
Kriegspiel	Avalon Hill	70
Kursk	SPI	71

Lankhmar	TSR	76
*Lee at Gettysburg**	SPI	73
Lee Moves North (See Lee at Gettysburg)	SPI	73
Lee vs Meade	Rand	74
Lee vs Meade (2d Ed)	Gamut of Games	74
Legion	SPI	75
*Leipzig**	Poultron (SPI)	69
Leipzig (2d Ed)	SPI	71
Lensman	Spartan	70
Leyte	SPI	75
Leyte Gulf	Avalon Hill	73
Ligny	SPI	76
Little Big Horn	TSR	76
Lost Battles	SPI	71
Luftwaffe	Avalon Hill	72
Lutzen (Thirty Years War)	SPI	76
Madame Guillotine	Gametesters	75
*Manassas**	Historical Simulations	75
Manassas (2d Ed)	GDW	76
March on India	Jagdpanther	75
Marching to Richmond	David Watts	75
Marengo	SPI	75
Marine	Jagdpanther	76
Marne	SPI	72
Marston Moor	Perfidious Albion	76
Mech War '77	SPI	75
Mercenary	Gametesters	75
Metamorphosis Alpha	TSR	76
Midway	Avalon Hill	64
*Minuteman**	SDC	72*
Minuteman	SPI	76
Missile Boat	Rand	74
Missile Crisis	AWA	76
Modern Battles Quadrigame	SPI	75
Monsters! Monsters!	Metagaming Concepts	76
Moscow Campaign	SPI	72
*Move Out**	ADA	71
*MP 44**	Jagdpanther	73
Mukden 1905	James Gabel	76
Mukden (Modern Battles Quad*)*	SPI	75
Musket & Pike	SPI	73
*Napoleon**	ADA	72
Napoleon	Gamma Two	74
*Napoleon at War**	ADA	74
Napoleon at War Quadrigame	SPI	75
Napoleon at Waterloo	SPI	71
Napoleon at Waterloo (Expansion)	SPI	72
Napoleon's Last Battles (Quadrigame)	SPI	76
Napoleon's Last Campaigns	Rand	74
Narvik	GDW	74
Nato	SPI	73
New Orleans, 1815	Challenge	76
Norad	SDC	73
Nordlingen (Thirty Years War)	SPI	76
*Normandy**	Poultron (SPI)	69
Normandy (2d Ed)	SPI	72
North Africa Quadrigame	SPI	76

*Nova**	Third Millennia	72
*Nowsville, 1980**	Jagdpanther	75
Nuclear Destruction	Flying Buffalo	76
Oil War	SPI	75
Oil Wars	David Watts	75
Okinawa	SPI	75
*Omaha Beach**	Spartan	72
Omaha Beach	Rand	74
*Operation Greif**	Guidon	70
*Operation Market Garden**	Third Millennia	72
Operation Olympic	SPI	74
Operation Rheinubung	Conflict Simulation	75
Origins of WW II	Avalon Hill	71
Outreach	SPI	76
Overlord, The Normandy Campaign	Conflict	73
Panzer '44	SPI	75
*Panzer Leader**	Avalon Hill	74
Panzer Leader (2d Ed)	Avalon Hill	75
PanzerArmee Afrika	SPI	73
*PanzerBlitz**	Avalon Hill	70
PanzerBlitz (2d Ed)	Avalon Hill	71
*Panzerdivision**	ADA	72
Panzergruppe Guderian	SPI	76
*Paris Commune, 1871**	Jagdpanther	74
Patrol	SPI	74
*PBI**	Third Millennia	72
Pearl Harbor	GDW	77
Peninsular War	David Watts	75
Pennine Wars	David Watts	75
*Phalanx**	SPI	71
*Pharsalus**	ADA	72
*Pilsudski**	Panzerfaust	69
*Poland 1939**	ADA	72
*Poland 1939**	Jagdpanther	74
Poland 1939 (2d Ed)	Jagdpanther	76
*Poltowa**	Panzerfaust	70
Port Arthur (Russo-Japanese War)	GDW	76
PQ-17	Jagdpanther	75
Princeton, 1777	AHIKS	74
Punic Wars	SPI	75
Quatre Bras (Napoleon's Last Battles)	SPI	76
Quebec, 1759	Gamma Two	72
Quest of the Magic Ring	Land of Legend	75
Raiders of the North	Taurus	75
Red Star/White Star	SPI	72
Redcoat	Empire	75
Remagen (Westwall)	SPI	76
*Renaissance of Infantry**	SPI	70
Revolt in the East	SPI	76
*Revolution**	ADA	71
Richthofen's War	Avalon Hill	73
Rifle & Saber	SPI	73
Rifle-Musket	SDC	74
Rift Trooper	AWA	76
Rigelian Wars	Jagdpanther	75
*Rising Sun**	Robert Partanen	70
Risk	Parker	64
*RN**	Jagdpanther	74

*Road to Richmond**	SPI	77
Rocroi (Thirty Years War)	SPI	76
Rommel	Loren Sperry	73
Rommel: War for North Africa	Rand	75
*Russian Campaign, The**	Jedko	74
Russian Campaign, The (2d Ed)	Avalon Hill	76
Russian Civil War	SPI	76
*Russian Front**	Spartan	71
Russo-Japanese War	GDW	76
*Sadowa**	ADA	73
Saipan	SPI	75
Salamanca	Maplay	76
*Salerno**	ADA	72
*Salerno**	Third Millennia	72
Saratoga 1777	Rand	74
Saratoga 1777 (2d Ed)	Gamut of Games	74
*Savo Island**	ADA	72
Schutztruppe	Jim Bumpas	75
*Sea Lion**	Third Millennia	73
Search and Destroy	SPI	74
Seastrike	Wargames Research	74
*Second Galactic War**	Third Millennia	73
Seelowe	SPI	74
Seven Days Battles	Battleline	73
Shenandoah	Battleline	75
*Shiloh**	Third Millennia	71
Shiloh	SPI	75
Siam	Potomac Games	76
*Sicily**	Drumco	70
*Sidi Rezegh**	Jagdpanther	73
Siege	Fact and Fantasy	74
*Siege of Barad-Dur**	Jagdpanther	75
Siege of Jerusalem, 70 AD	Historical Perspectives	76
Siege of Leningrad	Jagdpanther	76
Siege of Minas Tirith	Fact and Fantasy	75
Siege of Port Arthur	Strife	76
Sinai	SPI	73
*Sink the Bismarck**	Spartan	71
Sixth Fleet	SPI	75
Sniper	SPI	73
Soldiers	SPI	72
Solomons Campaign	SPI	73
Sorceror	SPI	75
Spacewar	TTA	72
*Spanish Civil War**	Jagdpanther	75
Spanish Civil War (2d Ed)	Jagdpanther	75
Spartan	SPI	75
*Spirit of '76**	Panzerfaust	72
Spitfire	SPI	73
SSN	GDW	75
*Stalingrad**	Avalon Hill	63
Stalingrad (2d Ed)	Avalon Hill	74
*Stalingrad III**	SPI	72
Stalk-1	CinC	76
Star Raider	AWA	75
Starfire	Eagle Games	75
StarForce	SPI	74
Starlord	Flying Buffalo	72

Starship	Fantasy	75
Starship and Empire	R-Squared Games	76
*Starship Combat**	Gary McGath	75
Starship Troopers	Avalon Hill	76
StarSoldier	SPI	77
Star Web	Flying Buffalo	76
Stellar Conquest	Metagaming Concepts	75
Strategy I	SPI	71
Strike Force One	SPI	76
Submarine	Battleline	76
Super Tank I	SATS	76
Supercharge (North Africa Quad*)*	SPI	76
*Suribachi**	Spartan	72
*Syria, 1941**	Abwehr	72
Tactical Game 3 (see *PanzerBlitz*)*	Poultron (SPI)	69
*Tactics II**	Avalon Hill	58
Tactics II (2d Ed)*	Avalon Hill	61
Tactics II (3d Ed)	Avalon Hill	72
Tank	SPI	74
*Tannenburg**	Poultron (SPI)	69
*Tannenburg**	ADA	72
*Tarawa**	Jagdpanther	73
Terrible Swift Sword	SPI	76
*Thapsus**	ADA	72
Their Finest Hour	GDW	76
Thermopylae	Kriegkit	73
*Third Reich, The Rise and Decline of the**	Avalon Hill	74
Third Reich (2d Ed)	Avalon Hill	76
Thirty Years War Quadrigame	SPI	76
Tobruk (Special Edition)*	Avalon Hill	75
Tobruk	Avalon Hill	75
Torgau	GDW	74
*Trafalgar**	Roger Cormier	68
Triplanetary	GDW	73
Trireme	Decalset	?
Tsushima (Russo-Japanese War)	GDW	76
Tunisia 1943	Excalibre	76
Turning Point	SPI	72
*Twelve O'Clock High**	Poultron (SPI)	70
Two Towers	Fantasy Games	75
*U-Boat**	Avalon Hill	59
UFO	Jeffrey Daugherty	75
Undersea Warriors	Taurus	75
Unentschieden	GDW	73
USN	SPI	71
Venerable Destruction	Excalibre	76
Verdun	Conflict	72
Vicksburg	Rand	75
*Viet Nam**	Gamescience	65
Viet Nam (2d Ed)*	Renwal	69
Viking	SPI	75
Viva!	Flying Buffalo	75
Von Manstein's Battles	Rand	75

Wacht Am Rhein	SPI	77
Wagram	SPI	75
*War at Sea**	Jedko	75
War at Sea (2d Ed)	Avalon Hill	76
War Between the States	SPI	77
War in Europe	SPI	76
*War in the East**	SPI	74
War in the East (2d Ed)	SPI	76
War in the West	SPI	76
War of the Ring	Fantasy	76
War of the Sky Galleons	Fact & Fantasy	76
War of the Worlds II	Rand	74
War of the Wizards	MAR Barker	75
War of 1812	Gamma Two	73
Warriors of the Dark Star	Taurus	75
Warsaw Pact	Jagdpanther	76
*Waterloo**	Avalon Hill	61
Waterloo (2d Ed)	Avalon Hill	75
Wavre	SPI	76
Wellington's Victory	SPI	76
Wellington in the Peninsula	Rand	76
Westwall Quadrigame	SPI	76
*White Bear, Red Moon**	Chaosium	75
White Bear, Red Moon (2d Ed)	Chaosium	76
Wild West	Usborne	76
Wilderness Campaign	SPI	72
William at Hastings	Falchion	76
William the Conqueror, 1066	TSR	76
Winter War	SPI	72
Wolfpack	SPI	74
*Wooden Ships & Iron Men**	Battleline	74
Wooden Ships & Iron Men (2d Ed)	Avalon Hill	75
World War I	SPI	75
World War II	SPI	73
World War II	Usborne	76
World War III	SPI	75
Wurtzburg (Modern Battles Quad*)*	SPI	75
Year of the Rat	SPI	72
Yeoman	SPI	75
Ythri, The	Metagaming Concepts	75
Zeppelin	Jagdpanther	74
*Zulu**	Combat	74
*1776**	Avalon Hill	74
1776 (2d Ed)	Avalon Hill	75
1776	Reiss	75
1812	SPI	72
1815: The Waterloo Campaign	GDW	76
*1914**	Avalon Hill	68
1914 (2d Ed)*	Avalon Hill	70
*1914 Revision**	Poultron (SPI)	69
*1918**	Poultron (SPI)	70
1918 (2d Ed)	SPI	72
1944	Histogames	73
4000 AD	House of Games	72
7th Cavalry	AWA	76

DIRECTORY OF PUBLISHERS

The following listing of wargame publishers is, at best, incomplete. The growth patterns of the hobby are such that many small publishers spring into and out of existence at a rate too fast for any book publishing schedule to pace accurately. If the reader is interested in any of the titles found in the Game Directory, he should write directly to the publisher for the most recent product literature available.

Avalon Hill Company	4517 Harford Rd., Baltimore, Md. 21214
Balboa Game Company	Box 989, Bellflower, Ca. 90706
Battleline Publications	P.O. Box 1064, Douglasville, Ga. 30134
Conflict Game Company	201 Broadway, Normal, Ill. 61761
CONSAD Computer Games	121 N. Highland Ave., Pittsburgh, Pa. 15206
Conquest Game Co.	1122 W. Burbank Blvd., Burbank, Ca. 91536
Excalibre Games	Box 29171, Brooklyn Center, Minn. 55429
Fact and Fantasy	P.O. Box 1472, Maryland Heights, Md. 63043
Flying Buffalo, Inc.	P.O. Box 1476, Scottsdale, Ariz. 85252
Game Designers Workshop	203 North St., Normal, Ill. 61761
Gamma Two Games, Ltd.	P.O. Box 46347, Vancouver, B.C., Can. V6R 4G6
GHQ Micro Armor/Miniatures	2634 Bryant Rd. South, Minneapolis, Minn. 55408
Guidon Games/Panzerfaust	P.O. Box 869, Fallbrook, Ca. 92028
Histo Games	34 Sharon St., Brooklyn, N.Y. 11211
Imperial Games	2305 West Park Row, Arlington, Tx. 76013
Loren Sperry Games	1014 N. 4th St., Sheboygan, Wisc. 53081
Martial Enterprises	P.O. Box 1315, National City, Ca. 92050
Metagaming Concepts	P.O. Box 15346, Austin, Texas 78761
Panzerschiffes	P.O. Box 2774, Akron, Ohio 44301
Rand Game Associates	Box 1776, Liberty Corners, N.J. 07983
Research Games, Inc.	200 Fifth Ave., New York, N.Y. 10010
Simulations Design Corporation	P.O. Box 19096, San Diego, Ca. 92119
SPI (Simulations Publications, Inc.)	44 East 23rd St., New York, N.Y. 10010
Taurus Games, Ltd.	1010 N. Lasalle St., Chicago, Ill. 60610
Third Millenia	1103 Whitworth St., Jackson, Miss. 39202
TSR Games	P.O. Box 756, Lake Geneva, Wisc. 53147
Zocchi Associates	1513 Newton Dr., Biloxi, Miss. 39532

SECTION VIII:

Symbols, Terms, and Format

by Redmond A. Simonsen

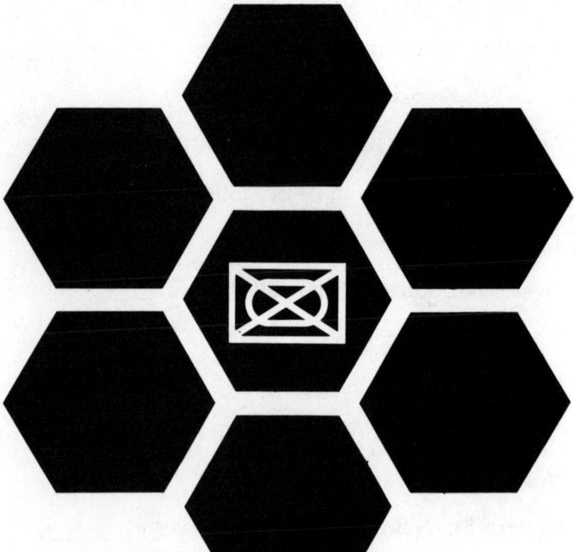

This part of the book might have been more succinctly titled "The Utility Section". In it I hope to give the reader the basics of wargaming terminology, symbology, and rules structure (as developed and used at SPI). As with most fledgling industries, there are few universal standards that are rigorously applied in the publishing of products of wargaming companies. In the past half-dozen years, I've been promoting the adoption of some of the standard styles developed by my colleagues in our effort to bring some sort of order to blizzard of conflicting usages and forms in gaming.

What is presented in this section is by no means complete and thorough-going. There are many more terms and formats that cry for regularization (at least in their general sense if not in detail). There are many advantages that would accrue to both the creators and the consumers of wargames if more standard forms were adopted. Not the least of these would be the reduction in learning time for both the gamer and the neophyte designer.

Meanwhile, because games are such bears to learn and digest I'd like to offer the following outline of a systematic approach to studying any new game one buys. It's worth one's while to use such an approach (though one may certainly develop a more personalized version of what I present here). Even the most "standard" seeming new game usually has some twist in its rules to undo the careless gamer. This can be particularly embarassing when, in the middle of a hotly contested game, one's opponent pulls a "trick" rule maneuver that decides the outcome.

An Approach to Learning Wargames

1. Spread out the map (and any loose chart sheets) on a large table. Scan this material without attempting to memorize it.

2. Briefly examine the sheet(s) of die-cut cardboard playing pieces (but don't punch them out yet).

3. Read the introductory material (usually found in the first few columns of the rules). If there is a parts inventory (as there is in SPI games) check off against it to make sure you *have* the whole thing (even the best of companies occasionally mis-packs their products).

4. Look at the Terrain Key and/or Terrain Effects Chart and familiarize yourself with the environment that you're going to be dealing with.

5. Examine the Combat Results Table(s) and keep in mind the value of the playing pieces.

6. Skim through the entire rules booklet (in SPI rules, simply read all the *General Rules* sections).

7. If there is more than one scenario (game situation) given, read the first one and read the victory conditions (i.e., the object of the game).

8. Carefully punch out the playing pieces and set them up according to the scenario instructions.

9. Now read *all* the rules (paying special attention to the Sequence of Play). As you come to a described procedure, try it out on the map (and then set the counters back into their starting positions). Use a pencil to underscore any rules that you think critical or any that you have difficulty understanding.

10. If you have an opponent available, play a trial game *together,* i.e., advising each other of any insights you may have regarding tactics or strategy for *both* sides. Cooperate — don't try to *win* the first game. If you *don't* have an opponent immediately available, play the game against *yourself.* It's very simple to do and it can actually be fun.

It will usually take about two to four playings to truly digest all the ins-and-outs of a game to the point where it becomes second nature. Of course, it's always time-saving to have a player already well-versed in the game teach it to you (but don't rely totally on his evaluations of proper strategy and tactics).

MILITARY UNIT SYMBOLS

The military symbol is a kind of graphic shorthand which permits virtually any type of military unit to be depicted in a compact easily recognizable form. In materials produced by Simulations Publications, they are most commonly found in organization diagrams, campaign maps, and on the playing pieces of conflict simulation games. The armed services of our country (and most foreign armed services) use them for much the same purposes.

The object of this data section is to brief our readers on the proper use of these symbols; to provide a comprehensive guide to their meaning; and to supplement the standard symbols with those that have special application in simulation games. The basic reference used to research this brief is United States Army Field Manual 21-30 (which contains more than you'll ever want to know about military symbols).

The Basic Symbol

Simple geometric shapes form the body of the basic symbols used to represent units, installations and activities.

1. A unit

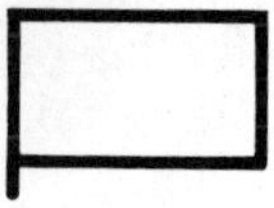

2. A headquarters or element of a headquarters

3. An observation post.

4. A logistical support unit (brigade-level trains and below)

5. An administrative or logistical installation

6. A logistical unit within a logistical chain of command.

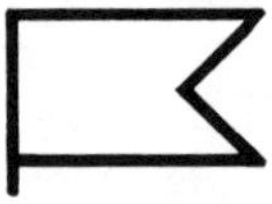

7. A logistical command headquarters within a logistical chain of command.

Developing the Basic Symbol

By placing other symbols within the basic shapes, specific types of military units can be described. A symbol denoting the size of the organization is placed on top of the basic shape, and the name (designation) of the unit is placed to the left of the basic shape. Two higher echelons of command can be noted to the right of the basic shape. Other information contributing to the identity of the unit may be placed directly below the basic shape (such as basic organic weapons or vehicles).

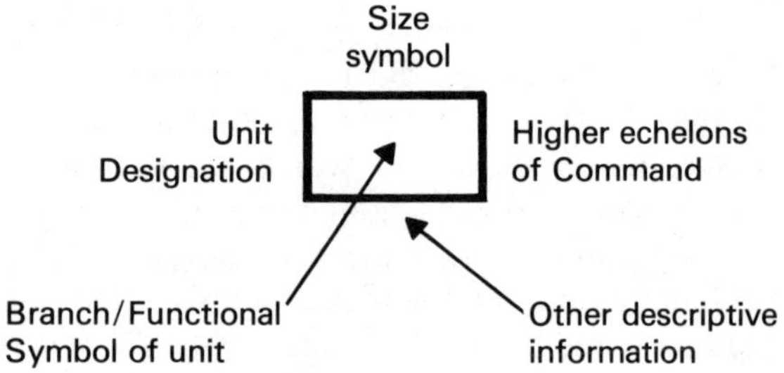

The Symbols Used to Describe Unit Types

There exists a whole "vocabulary" of symbols which when used independently or in combination can describe virtually any unit-type. These symbols fall into two main groups.

1. *Branch symbols* (which in themselves stand for the various major branches within the Army);

2. *Functional symbols* (role or environment-describing symbols).

Branch symbols can be used independently, or in combination with other branch symbols or functional symbols. Functional symbols are rarely used independently. Note that those branch and functional symbols which have little application to simulation usage are not included in this brief (e.g., Finance, Data Processing, Topographic, etc).

BRANCH SYMBOLS

1. Air Defense

2. Armor

3. Chemical

4. Coastal Artillery

5. Cavalry/Recon

6. Engineer

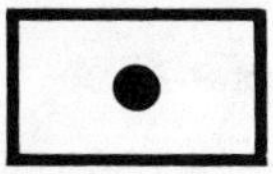

7. Field Artillery

8. Infantry

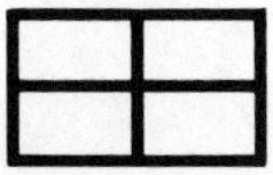

9. Medical

10. Military Intelligence

11. Military Government

12. Military Police

13. Ordnance

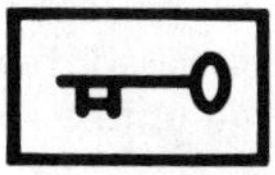

14. Quartermaster

15. Signal

16. Transportation

17. Veterinary-remount

FUNCTIONAL SYMBOLS

18. Airborne

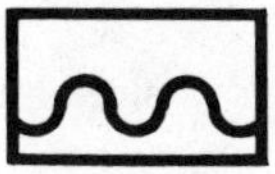

19. Amphibious

20. Antitank

21. Army Aviation

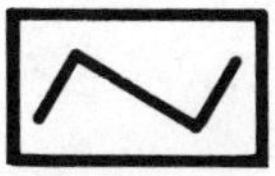

22. Electronic Warfare

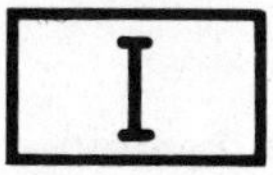

23. Irregular Forces

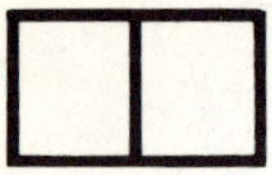
24. Temporarily Motorized

25. Motorized (cross-country)

26. Mountain

27. Parachute (jump-qualified; not assigned to airborne unit)

28. Psychological Warfare

29. Repair and Maintenance

30. Replacement

31. Rocket/Missile

32. Supply

33. Airmobile (organic to airmobile organizations)

34. Airmobile (unit possesses the aircraft to perform airmobile operations)

Branch and Functional Symbols Combined

The preceding symbols can be combined (sometimes using additional specialized symbols) to create a whole range of unit symbols:

35. Air Cavalry

36. Airborne Armor

37. Airborne Artillery

38. Airborne Infantry

39. Airborne Medical

40. Airborne Reconnaissance

41. Airborne Signal

42. Airmobile Artillery

43. Airmobile Engineer

44. Airmobile Helicopter

45. Airmobile Infantry (aircraft possessing)

46. Airmobile Infantry

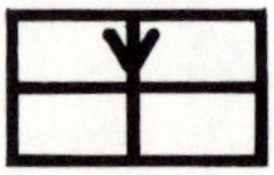
47. Airmobile Medical

48. Amphibious Armor

49. Antitank Artillery

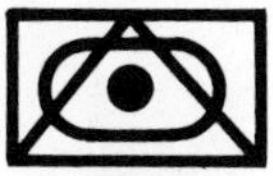
50. Armored Antitank

51. Armored Artillery

52. Armored Cavalry Recon

53. Armored Car

54. Bridging Engineers

55. Horse Artillery

56. Infantry Artillery

57. Mechanized (Armor) Engineer

58. Mechanized (Armor) Infantry

59. Mechanized (Armor) Infantry Artillery

60. Mountain Infantry

61. Supply and Transport

Non-Standard Symbols

The following symbols are not to be found in FM 21-30, but rather have been developed by SPI for use in special applications.

62. Marine or Naval Infantry

63. Machine gun unit

64. Motorcycle recon

65. Ranger or Commando

World War Two Era Symbology

The following symbols (some of which are antecedents of contemporary symbols) will sometimes be found in use in SPI material.

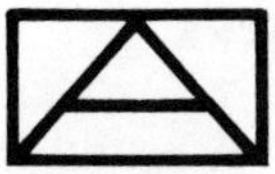
66. Antiaircraft Artillery

67. Armored Train

68. Glider

69. Guerrilla

70. Heavy Weapons Infantry

71. Headquarters

72. RR Gun

73. Rocket Artillery

Organizational Size Symbols

In order to indicate the size of the unit being depicted, the following symbols are placed on top of the basic symbols. Note that the organizations in parentheses are the approximate U.S. Air Force organizational equivalents.

Squad	•
Section	••
Platoon (Section)	•••
Company, Battery, Troop, (Flight)	I
Battalion (Squadron)	II
Regiment (Group)	III
Brigade	X
Division (Wing)	XX
Corps (Air Division)	XXX
Air (Numbered Airforce)	XXXX
Army Group (Major Air Command)	XXXXX
Theater of Operations —not in FM 21-30—	XXXXXX

When it is necessary to indicate that a unit is markedly understrength due to losses or detachments, a minus sign in parentheses will appear to the right of the unit symbol:

(–)

When it is necessary to indicate that a unit has been substantially reinforced, a plus sign parentheses will appear to the right of the unit symbol:

(+)

A task force (temporary grouping of units under one command) will be indicated by an upside down "U" shaped bracket over the approximate size symbol:

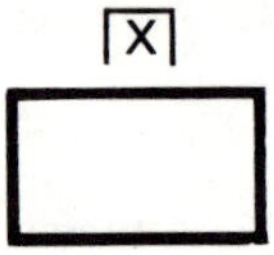

A battlegroup (or Kampfgruppe) will be indicated by the replacement of the usual size symbol with the abbreviation BG (or KG). Note this is an SPI usage, not in accordance with FM 21-30.

BG

A cadre (i.e. the experienced personnel remnants of a destroyed or disbanded unit) will be indicated by the replacement of the usual size symbol with the abbreviation CDR (SPI usage).

CDR

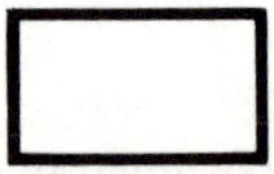

Examples of Symbols in Use

The 1st Battalion of the 3rd Brigade/42nd Infantry Division:

II

1  3/42

The 3rd Brigade of the 42nd Infantry Division/5th Infantry Corps:

X

3  42/5

The 42nd Infantry Division of the 5th Corps/8th Army:

Note that the above usage is a mixture of FM 21-30 usage and a simplified SPI usage.

Logistical Installations and Activities

Only the main classes of supply installations are shown (there are many other, more specific variations).

74. All Classes

75. Class I — Subsistence

76. Class II — Clothing, "Housekeeping" materiel

77. Class III POL (Petrol, Oil, Lubricants)

78. Class IV — Construction

79. Class V — Ammunition

80. Class VI — Personal Demand Items

81. Class VII — Major End Items

82. Class VIII — Medical Supplies

83. Class IX — Repair Parts

MISCELLANEOUS INSTALLATIONS

84. Hospital or Aid Station

85. Prisoner of War Collection Point

86. Traffic Control

87. Water

Aviation Symbols

The following aviation symbols describe Air Force units. By making the propeller component of the symbol *solid*, similar Army Aviation units may be symbolized. Note, certain of these symbols are interpolations designed by the author (not in accordance with FM 21-30) and are marked with an asterisk (*).

88. Any Air Force Unit

89. Fighter

90. All Weather Fighter

91. Fighter Bomber

92. Tactical Strike Fighter (Light)

93. Bomber

94. Medium Bomber*

95. Heavy Bomber*

96. Reconnaissance

97. Light Fixed Wing

98. Drone

99. Rescue

100. Anti-Submarine

101. Seaplane

102. Transport

103. Helicopter

104. Helicopter Transport

105. Med Evac Helicopter*

106. Attack Helicopter*

Tactical Weapons Symbols

The following weapons symbols can be used independently on tactical maps and units or in conjuction with unit symbols (when placed below them) to indicate the main organic weaponry of a unit.

The system is built upon the use of only two basic symbols which are then modified in meaning by the application of a few other symbols to denote approximate size, general characteristics and the role the weapon is being used in (i.e., either a ground combat role, an anti-tank role, or an anti-aircraft role). To depict a given weapon the following procedure is used:

Select one of the two basic weapon symbols

Basic
Infantry
Weapon

Basic
Artillery
Weapon

Add one (for "medium") or two (for "heavy") horizontal bars to denote the approximate size of the weapon. When the weapon is in the "light" category, no horizontal bars are added.

Examples:
Light
Automatic
Weapon

Medium
Artillery
Gun

If the weapon is one with a high trajectory, a circle is added to the base of the shaft:

Example:
Medium
Mortar

If the weapon is a flat trajectory Weapon in an anti-tank role an inverted "V" is added to the base of the shaft.

Example:
Light
Anti-tank Gun

If the weapon is primarily designed for an airdefense role, a closed semi-circle is placed at the bottom of the shaft:

Example:
Heavy
Anti-aircraft
Gun

If the weapon is a rocket-projector or launcher a double inverted "V" is placed at the head of the shaft:

Light Artillery
Rocket Launcher

If the weapon is mounted on a fully-tracked self-propelled chassis, a diamond is placed below the weapon-symbol.

Example:
Tracked
Self-Propelled
Medium Howitzer

If the weapon is mounted on a wheeled, self-propelled chassis, a horizontal bar with two circles and a diamond is placed below the symbol.

Example:
Self-propelled
Light
Anti-aircraft
Gun

The Table of Symbols gives a comprehensive breakout of weapons in the three size categories.

Table of Weapon Symbols

Light Medium Heavy

Automatic Infantry Weapon

Mortar

Anti-aircraft Machine Gun

Artillery Gun or Gun/Howitzer

Howitzer

Anti-tank Rocket Launcher

Anti-tank Gun

Recoilless Rifle

Rocket Launcher Artillery

Anti-aircraft Gun

Missile

Air Defense Missile

Anti-tank Missile

Surface-to-Surface Missile [SSM] Launching Site

Surface-to-Air Missile [SAM] Launching Site

GAMESPEAK: A Glossary of Terms Used in Wargaming

What this glossary attempts to do is standardize as much of the jargon as possible to reduce the learning load upon all of us and to inject more consistency and logic into our game rules (and, one hopes, into those of our brother publishers — who should feel free to use any and all of SPI's terminology). The glossary is *not* exhaustive (although it is, perhaps, exhausting). Many more pages could be devoted to all the fine turns of phrase inflicted on the wargaming world by this writer and others.

advance (after combat). A move made as a result of successful combat

Administrative Capacity. A numerical rating of the administrative ability of a leader or general staff.

Air Attack Strength. The numerical rating of the ability of a unit to attack air units.

Air Combat Strength. The numerical rating of the ability of a unit to either attack or defend against air units.

Air Combat Phase. That part of the Game-Turn during which air combat takes place.

Air Defense Strength. The numerical ability of a unit to defend against air units.

air defense unit. A unit, the primary mission of which is to provide air defense; in WWII, a flak unit.

air interdiction. An action taken by air units (or abstract air capability) to interfere with Enemy movement and/or supply.

Air Interdiction Phase. That part of the Game-Turn during which air interdiction may take place.

Air Movement Phase. That part of the Game-Turn during which the movement of air units takes place.

airlift. The transporting of ground units via air.

Airlift Capacity. The capability of a Player's forces to airlift a given number of units or Strength Points in a given period of time. This capacity is sometimes expressed in terms of Airlift Points.

Air Range Allowance. The per-Phase one-way movement limit of an air unit expressed in terms of hexes; in effect, the radius of action of the air unit.

Air Movement Allowance. The total per-Phase movement limit of an air unit expressed in terms of hexes.

airborne unit. A military unit capable of parachuting or glider landing.

air landing unit. A military unit capable of air landings in hostile territory.

airmobile unit. A military unit capable of heliborne operations.

air supply. The technique of supplying ground units via airdropped or delivered supply.

Ammunition Point. A modular quantity of ammunition used by artillery or other types of units.

anti-aircraft unit. See air defense unit.

anti-tank unit. A unit, the primary mission of which is to defend against Enemy armor.

anti-submarine unit (ASW unit). A unit, the primary mission of which is to defend against and destroy Enemy submarines.

armored unit. A unit, the main strength of which is composed of tanks or other armored fighting vehicles (AFV's).

army. A military unit consisting of two or more corps; or the whole of a nation's ground strength.

artillery unit. A unit, the primary strength of which is composed of field guns or howitzers.

attack. A specific combat action against a particular Enemy unit or units.

attacker. The Player (or units of that Player) executing an attack.

Attack Strength. The numerical rating of the ability of a unit to perform offensively; expressed in Attack Strength Points.

Attack Supply. The specific supply condition that enables a given unit to make an attack.

attrition. A graduated loss process whereby a unit loses its strength bit by bit.

Attrition Phase. That part of a Game-Turn during which units automatically or probabalistically experience losses as a result of being out of supply, in a poor morale state or for some other similar reason.

autonomous force. A unit or collection of units that are not under the control of any Player, but rather which perform actions independently according to a set of rules.

battalion. A military unit consisting of two or more companies and a headquarters.

battery. A company-size artillery unit.

Blast Radius. The circular area, measured in hexagons from the point of impact, affected by the burst of an artillery shell or other heavy munition.

Blast Strength. A numerical rating of the effectiveness of an explosion in a given hex; if the explosion affects more than one hex, the strength usually declines logarithmically as one tracks outward from the center of impact.

bombardment. The systematic delivery of heavy munitions upon a specified target; the technique whereby artillery units may attack adjacent or non-adjacent Enemy units using their Bombardment Strength.

Bombardment Strength. The numerical rating of the ability of an artillery (or other) unit to attack using bombardment techniques.

break-down. The act of substituting two or more smaller component units for one larger unit.

brigade. A military unit consisting of two or more battalions and a headquarters.

case. A numbered or lettered paragraph that explains or sets forth a specific detail of a major rule.

Casualty Points. A numerical measure of losses experienced by an army. Casualty Points may or may not be directly related to the Combat Strength of the actual units lost.

cavalry units. A mounted unit that fights on horseback. In modern usage, a light reconnaissance unit mounted in armored cars or scout vehicles.

Close Assault. In tactical games, a special attack made by infantry or infantry/engineers.

column. A unit formed in marching order or a collection of units traveling single-file on a road. In some usages, a battle formation that attacks presenting a minumum front to the enemy.

conflict simulation. A game which attempts to model some of the aspects of a struggle between two or more opposing forces. Used to refer to a body of games broader than wargames (which are a sub-class of conflict simulations).

combat (or combat action). The interaction between two or more opposing units that results in losses, retreats or other changes of state.

Combat Phase. That part of the Game-Turn during which combat takes place.

Combat Strength. The numerical rating of the ability of a unit to attack and defend; expressed in terms of Combat Strength Points (or simply Strength Points in games where the values are not split into two or more characterizations such as Attack Strength and Defense Strength).

Combat Results Table. A table that sets forth the possible results of all combats allowed within the game system. Usually the table is probabalistic, i.e., the greater the Strength brought to bear the higher the chance for success based upon the throw of a die or some other random number generation device.

combat results (or outcomes). The specific possible effects of combat within the game-system; also, the descriptive labels used to identify those effects. What follows is a list of some of the more frequently used combat results and their general meaning:

Attacking Units Eliminated [Ae]: all of the participating attacking units are destroyed (removed from play).

Attacking Units Retreat [Ar]: all of the participating attacking units must retreat one or more hexes.

Attacking Units Disrupted [Ad]: all of the participating units are disrupted (which usually means they are pinned in place or suffer some other change of state).

Attacking Units Lose Indicated Number of Steps or Strength Points [A1, A2, etc.]: one or all of the participating attacking units are reduced by the number indicated (or retreat the indicated number of hexes).

Attacker Parity Elimination [APe]: The attacker loses participating units whose face value is at least equal to the face-value of the defending force (the defending force loses nothing.) Note that this is a new coinage to replace "Attacker Exchange" — a less accurate title for the same result.

One-Half [or other fraction] Attacker Parity Elimination [½APe]: The attacker loses participating units the face-value of which is at least equivalent to half that of the defending force.

Exchange [Ex]: The weaker force is destroyed and the stronger force must lose participating units the face-value of which is at least equal to that of the weaker force. Similarly, "2Ex" or "4Ex" etc.

Engaged [Eg]: The forces involved remain locked in combat (sometimes with further combat mandatory; in other systems mutual loss-taking is suffered, etc.).

Contact [Cn]: The forces involved had "discovered" each other and (usually) suffer no deleterious effect.

Counterattack [Ca]: The defending forces must immediately make an attack against the attacking force.

Pinned [Pn] or **Suppressed [Sp]:** The suffering forces may not move (or are in some other way temporarily inhibited). Used in tactical games; similar to disruption.

Panicked [Pk]: The suffering forces panic and (usually) scurry off out of control of the owning Player.

Routed [Rt]: The affected forces execute a headlong retreat; less intense than panic — more severe than disruption.

Defending Units Eliminated [De]: This and all the other analogous "Attacker" results previously specified can be applied to the defender simply by replacing the modifying prefix "A" with "D"

Readers should note that there are many more results and combinations of results possible in conflict simulations game systems. Those specified, however, are the most commonly experienced.

Combat Supply. The specific supply condition that enables a unit to attack and defend at normal strength.

combat unit. A general class of units distinguished by the ability to independently move and engage in combat.

command control. Units in command control move and engage in combat normally; units that suffer a loss of command control are restricted in one or both of these activities.

communications, line-of-. A line of connected hexes, free of Enemy interference, that can be traced from a given unit to some designated point on the map.

company. A military unit consisting of two or more platoons and a headquarters.

consolidate. The act of substituting one larger unit for two or more smaller, component units.

continuous line. An unbroken line of Friendly units and/or hexes controlled by Friendly units.

controlled hex. A hex upon which a unit is exerting an effective Zone of Control.

corps. A military unit consisting of two or more divisions, a headquarters, and attachments.

covering terrain. Terrain which provides concealment and/or protection from Enemy observation and fire (used in tactical games).

defender. The Player (or units belonging to that Player) who is the object of an attack.

Defense Strength. The numerical rating of the ability of a unit to perform defensively; expressed in Defense Strength Points.

debark. The off-loading of units from ships; similarly deplane, detrain. Antonyms: embark, emplane, entrain.

demoralization. The condition suffered by an army (or units of that army) that has experienced heavy losses. Usually affects movement and combat performance.

Demoralization Level (or **Threshold).** The point at which an army or units of that army become demoralized; usually measured in terms of strength lost.

depot unit. A military unit or station from which combat units draw supply.

depletion. A unit state connoting the exhaustion of some critical element of materiel and/or personnel from which they may recover by undergoing some sort of re-building process.

designation. The official title of a specific military unit, e.g., the 42nd Infantry Division.

differential. The difference in strength between an attacking force and a defending force. This number is often used to head the results columns on a Combat Results Table.

direct fire. In tactical games, flat-trajectory fire along the line-of-sight using crew-sighted weapons.

dismount. The off-loading of troops from vehicles or animal carriers. Antonym: mount.

diversionary attack. An attack performed primarily in order that a related attack might have a better chance of success. Usually done in games that require adjacent units to attack.

division. A military unit consisting of two or more brigades (or regiments) plus a headquarters and attachments.

Double-match combat. A combat in which the attacker gains special advantage by having a force twice the strength of the defender.

dummy unit. A counter that resembles an actual unit but which has only the function of deceiving the Enemy Player.

ECM. Electronic Counter-measures. The technique of using jamming devices and decoys to confuse Enemy radar and other sensing devices.

Effectiveness Rating. The numerical rating of the ability of a unit to perform in a combat situation; usually used in conjunction with some other value.

Enemy. The opposing Player or units of the opposing Player.

engineer unit. A military unit the primary function of which is to aid other units in the clearing of minefields, crossing of rivers, etc., under combat conditions.

entry cost. The Movement Point cost to enter a given type of hex.

environment. The physical spheres in which the game takes place, i.e., land, air, or sea.

exit-cost. The Movement Point cost to leave a given hex.

face-value. The strength(s) or value(s) printed

on a given counter (as opposed to the effective strength in a given situation).

facing. The direction in which a unit is pointed (either to move or to fire its weapons). Used in tactical games where units have a "front" and "flanks."

field of fire. The arc of space through which a unit may direct the effect of its weapons (e.g., the field of fire of a machine gun).

Fire Strength. The numerical rating of the missile-firing ability of a unit, e.g., an artillery unit or a crossbow unit.

First Player. The Player that takes action first in the sequence of play.

fort or fortification. An emplacement printed on the map or represented by a counter, that augments the Defense Strength of units in it.

free deployment. An initial placement of units that allows the Player greater discretion than the historical positioning.

game map. The playing surface.

Game-Turn. See Sequence of Play.

General Rule. A major game rule expressed in a brief, general paragraph.

General Supply. That specific supply condition that enables a unit to maintain itself on a day-to-day, "housekeeping" basis (as contrasted to the extraordinary supply required for attack).

grid. The regular pattern of lines used to position units on a game-map (i.e., in most games, the pattern of hexagons).

grain. The direction in which the straight rows of hexagons appear to run parallel to one or another edges of the map; "grain-short" means through the short dimension of the map and "grain-long" through the long dimension. Most maps are grain-short.

ground support. The use of tactical aircraft to aid the attack of Friendly ground units.

hex. A single hexagon in a hexagonal grid.

hex-side. One of the six sides of a hexagon.

hex number. In SPI games, a four digit number that describes (and is actually printed in) a specific hex.

holding area. A box printed on the map that is designed to contain units in some inactive or semi-active state.

indirect fire. In tactical games, high-trajectory fire guided by an observer; fire not restricted by line-of-sight.

interdict. To interfere with the Enemy line-of-communications (usually by the use of airpower or artillery).

involved unit. A unit which is a direct participant in a given game action.

infantry unit. A unit, the main strength of which is composed of troops bearing individual weapons.

line-of-sight. The path of vision between two locations on a game-map.

mechanized unit. A military unit, the main components of which are mounted in armored units and mechanized infantry units.

melee combat. Hand-to-hand fighting (usually involving the use of edged weapons).

mountain infantry unit. A military unit specially trained and equipped to operate in mountainous terrain.

Movement Allowance. The numerical expression of the ability of a unit to move. In most games, a unit expends one Movement Point of its Movement Allowance in order to enter the most common hex type.

order of battle. The make-up of a Player's forces by organization and type; the forces available to a Player in a given scenario.

organic. In military parlance, a weapon or formation which is a formal part of the authorized make-up of a unit.

overrun. An overwhelming attack (usually executed as a unit moves) that enables a defensive position to be penetrated.

owning Player. The Player to whom a particular unit or geographic feature belongs.

passenger. A unit that is being transported by another unit.

Phasing Player. The Player whose phase it is at that particular point in play.

Platoon. A military unit composed of two or more squads.

Player. One of the active participants in a game.

plot. A detailed written description of an activity about to be executed in a simultaneous Phase.

Point. The unit-of-measure used in describing the constituent parts of a Movement Allowance, a Combat Strength, or any other similar value used in conflict simulations.

Production Center. A counter or installation printed on the map that is used to produce new units or that has some similar function.

Protection Strength. A Defense Strength which is inherent in a terrain type. The Defense Strength of units with respect to missile fire when other types of Defense Strengths are employed in the same system.

Rail Capacity. The per Game-Turn ability of a Player's rail net to move units and/or supplies by rail; measured in Rail Capacity Points.

Randomizer. A system of numbered chits used to generate (by drawing them from a "hat") a random number in a fashion similar to the use of a die or dice.

Range Allowance. The radius of action or effect (measured in hexagons) of an air unit, or the fire of an artillery or other missile unit.

reinforcement. A unit or force which does not begin the game in play, but rather which appears at some scheduled time later in the game.

Reinforcement Track (or **Schedule).** The schedule of reinforcements that a Player is authorized to receive in a given game.

retreat priority. An ordered listing of hex-types or hex-conditions into which a unit should retreat when called upon to do so as a result of combat.

road movement rate. The number of Movement Points per hex that a unit expends when moving by road.

scale. Literally, the number of miles or kilometers represented by the width of a single hex on the game map. More loosely, a game will be characterized as being a certain scale depending upon the size of the military organization represented by a single playing piece. In this sense there are three main scales:

Tactical. Single man, ship, vehicle or aircraft. Also team, squad, platoon, or company size units.

Operational. Battalion through division size units.

Strategic. Division through Army and Army Group size units.

Sometimes the above scales may be modified by the adjectives "low" or "high" to indicate a scale at one of the extremes of the basic category.

scenario. An organized description of the orders of battle, stating deployments and victory objectives that the Players will use to play a specific situation presented in a game.

Sequence of Play. The strictly defined order in which specific Players are to take specific actions in a standard Game-Turn. In most games, the parts of the Game-Turn (which is the basic unit of time in a game) are labelled using a hierarchy of standard names for each part.

What follows is a list of these standard parts, given in descending order.

Game-Turn, Player-Turn, Phase, Segment, Step.

In some games, there exists the need to group a series of Game-Turns into a repeatable set. If this set does not correspond to a calendar designation such as Game-Month or Game-Season, then the term *Game-Cycle* will be used. In other instances, there is the need to group a number of Player-Turns within a Game-Turn. In such cases the term *Stage* will be used. In some cases where two or more Players perform the same function at the same time, the appropriate part of the Game-Turn will be modified by the adjective *Joint.*

Squad. A military unit composed of nine to sixteen men.

stack. A collection of Friendly units occupying the same hex. The number of units that may be placed in the same hex is usually limited.

Stacking value. A numerical representation of the amount of "space" that a given unit occupies in a given hex expressed in terms of Stacking Points. In such a rule system, a hex is said to have a stacking limit (in Points) and the Player may place any number of units in the hex whose combined Stacking Values do not exceed that limit.

Standard Game. The body of rules which constitute the basic game (without any options or modifications).

Strength Step. A specified portion of the Strength of a unit which can be lost due to combat. In most systems the loss of a Strength Step requires that a new, weaker version of the unit be substituted for the counter in play.

supply. The necessary consumables used by a unit in movement and combat. A unit that has supply is said to be "in supply;" a unit that does not is "out of supply."

supply line. A path of connected hexes free of Enemy interference traced from a unit to its source of supply. Usually a supply line is cut when Enemy Zones of Control and/or units interfere with all the pathways to a Friendly unit.

supply unit. A mobile unit, the primary function of which is to provide supply for units in the field.

target acquisition. The fulfillment of sighting and other requirements necessary to be able to fire upon a given unit or hex.

target hex. The hex occupied by a unit that is the object of an attack (usually used in tactical games).

Task. A specific action that a unit may be assigned to perform at some specific time in the Game-Turn.

Task Allowance. A numerical rating of the ability of a unit to perform Tasks in the Game-Turn. A unit will expend a certain number of Task Points per Task.

Task plot. See plot.

terrain. The dominant physical characteristic of a hex (either natural or man-made).The following is a list of standard terrain nomenclature arranged in ascending order of effect upon movement and/or combat:

clear, mixed, sand, woods (tactical), forest (operational), swamp (or marsh), broken, rough, mountainous, impassable.

Bodies of Water:creek, stream, river (minor or major), estuary, sea or lake, ocean.

Terrain Effects Chart. A chart that details the interrelationships of terrain and unit-types with regard to effects on movement and combat.

unit. A playing piece which represents a military organization such as a division or a squad.

victory condition. The goal which must be

achieved by a Player in order to be considered the winner of the game.

victory level. The degree of victory attained. The standard levels of victory in ascending order are; Draw, Marginal Victory, Substantive Victory, Decisive Victory, Overwhelming Victory.

Victory Point. The unit-of-measure of victory in those game-systems that assign point values to objectives.

Zone of Control. The area of effect surrounding a unit; usually defined as the six immediately adjacent hexes. In theory, the exact character of a Zone of Control in a given game-system may be delineated by the use of a combination of adjectives, taking one from each of the following groups:

Effect on Movement:

Locking. Units must stop upon entering an Enemy controlled hex and may leave only as a result of combat.

Rigid. Units must stop upon entering an Enemy controlled hex and may leave only at the beginning of a Movement Phase.

Elastic. Units may enter and leave Enemy Zones of Control by paying an additional cost in Movement Points.

Open. Zones of Control have no effect on Movement.

Effect on Combat:

Active. Requires that *every* Enemy unit in a Friendly Zone of Control be attacked in the Combat Phase.

Inactive. Makes no requirement for attack.

Effect on Supply Line and Retreat:

Interdicting. Prohibits the path of retreat or supply from being traced through an Enemy Controlled hex regardless of the presence of Friendly units.

Suppressive. Prohibits the path of supply or retreat from being traced through an Enemy controlled hex if the hex is not occupied by Friendly units.

Permissive. Does not affect the path of supply or retreat in any way.

Example. A Rigid, Active, Suppressive Zone of Control. Note that if units have an Open, Inactive, Permissive Zone of Control they in fact have **no** Zone of Control.

(1.0) SPI RULES MANUSCRIPT STANDARDS

GENERAL RULE:

All rules manuscripts will conform to the specifications set forth in this style memo and any attachments to it. All manuscripts will conform to the specifications given on the SPI Manuscript Form and, furthermore, all manuscripts will be typed on the Manuscript Form paper. Some of the specifications may seem arbitrary or picayune. Nevertheless, there are good reasons for them and they will be followed. Not the least of these reasons is simple uniformity and consistency in support of cleaner and more coherent rules. Note that this memo itself is typed in rules style as an overall example of the format.

CASES:

(1.1) MAIN SECTION TITLING AND THE GENERAL RULE

(1.11) <u>Main Section Titling</u>: Rules should be broken down into titled Main Sections, each of which deals with a major rules concept or mechanism. The title should express the content, but should not be overly long; six words or less is preferable.

(1.12) <u>Main Section Ordering</u>: Main Sections should be arranged in basic conformance to the Sequence of Play of the game. In certain exceptional cases, a Main Section will be placed out of this natural sequence due to its specialized nature. if you decide to do this (and if there is any room for doubt), ask the Art Director for a ruling.

(1.13) The General Rule: The General Rule should sum up the section and touch upon every major thought in it. It should be no more than twenty (20) typewritten lines in length. If you find that you need more than this to explain the basic concepts, it is an indication that your Main Section should be broken up into two or more Main Sections (or that you are running off at the mouth in your General Rule). A General Rule should always be tagged as the "GENERAL RULE" as per the style set forth in this memo.

(1.2) PRIMARY CASES

Primary Cases (those numbered to the first decimal place) should always have titles. The title should be expressive of the basic content of the Primary Case and the Secondary Cases it comprises. The title should be no more than six words long (if at all possible) and should be independently identifiable (by that title) as belonging to that Main Section.
EXAMPLE: ZONE OF CONTROL EFFECT ON COMBAT

(1.21) Note that text may run under the Primary Case title only; or text may run under the Primary Case title and then be further described in Secondary Cases: or the Primary Case may consist of a title only, with all the explanatory matter contained in the Secondary Cases under that Primary Case.

(1.22) If the Primary Case has text running under its title, that text should be no more than sixteen (16) manuscript lines in length (and preferably twelve or less).

(1.3) SECONDARY CASES

A Primary Case will be broken out into two or more Secondary Cases when there are several specific points to be made under that heading. If a Primary Case runs more than twelve (12) manuscript lines, it is a good indication that the Case should be broken out into Secondary Cases.

(1.31) Length of Secondary Cases: Secondary Cases should be no more than sixteen (16) manuscript lines in length, and preferably twelve (12) or less. If you find that your Secondary Cases are consistently more than twelve lines, and that you usually have only three or fewer Secondary Cases under each Primary Case, then this is an indication that you should break up the information into smaller, more digestible bits.

(1.32) Titling Secondary Cases: Secondary Cases need not be titled. If the writer does wish to use titles, however, then all the Secondary Cases in that Primary Case must be titled. Secondary Cases have "run in" titles underscored as shown in this Secondary Case. Note also, that the title is in upper and lower case typing.

(1.4) EXAMPLES OF PLAY AND PROCEDURE
Whenever a procedure or play system is described it will be

followed by a specific example(s) of that procedure as used in the game. This can be a brief written example (or, if necessary, a diagram with supporting copy).

(1.41) Examples which are part of the case should be placed at the end of the case and set off by the word "EXAMPLE" beginning on a new line without skipping any space.
EXAMPLE: This is an example of an example.

(1.42) Examples can also be independent Secondary Cases (if the procedure being described is extensive).

(1.43) Illustrated examples should carry the illustration on a separate manuscript page, coded to correspond to the text page. When such an example is given, always signal its insertion on the text page by typing a key line enclosed in double parentheses:
EXAMPLE: ((insert Zone of Control diagram))

(1.5) CHARTS AND TABLES
Charts and Tables will be allocated an independent Main Section occuring just before the Scenarios (or if there are no scenarios, at the end of the rules before any optional rules and/or credits).

(1.51) Charts and tables will be referenced to and from the appropriate rules sections. Any lengthy explanations of tables will be incorporated into the appropriate Main Section as a

Primary Case bearing the title of the table followed or preceded by a modifying clause:
EXAMPLE: (6.8) HOW TO USE THE COMBAT RESULTS TABLE

(1.52) A tabular arrangement will be called a "table" when it consists mainly of numbers and/or when it is a probability table of any kind.

(1.53) A tabular arrangement will be called a "chart" when it consists mainly of columns of words not involving probabilities and/or when illustrations are included in it.

(1.54) A tabular arrangement will be called a "track" when it is used in conjunction with a marker to record points scored or the passage of time or some other recording function.

(1.55) A tabular arrangement will be called a "schedule" when it is basically a list of events or items and does not serve as a recording device or a probability table. For example, a trackless reinforcement list would be titled "Reinforcement Schedule" not "Reinforcement Track."

(1.56) Each chart or table will be typed on a separate page, appropriately numbered and referenced. Explanatory text for probability tables will not be typed on the same page, as the table (see 1.51).

(1.57) The standard results abbreviations found in the glossary will be used in all probability tables (results not covered by the abbreviations listed will be arrived at in consultation with the Art Director). Note that in tables which always affect only the Defender, the "A" and "D" prefix letters are dropped and the result code alone is used (capitalized).
EXAMPLE: De becomes E.
Numerical results that apply only to one Player throughout the table can be given as a simple, unmodified number.

(1.58) All tables will be typed with a generous amount of column space (the paper may be turned sideways for wide tables).

(1.6) SCENARIOS

Scenarios should be a Main Rules Section (or more than one Main Section if they are numerous and can be logically divided into major groupings). The Scenarios should be introduced by a General Rule which describes the organization of the information given in the Scenarios and a Procedure section describing how to set-up and play a game.

(1.61) If there are numerous scenarios, the Orders of Battle should be given in paragraph form rather than in columnar form. Starting hex numbers should be given in parentheses following the units:
EXAMPLE: 3i, 2m, 4a (4527); 2m (4528); 9c (4833).
EXAMPLE: two 4-4, three 4-6 (3107).

(1.62) If there are only a few scenarios and the set-up is fairly involved, give the initial set-up in columnar form with the hex number in the left-hand column:
EXAMPLE:
Starting Units

Hex	
1010	Three 4-2, one 8-4.
1118	Twenty-one AP, one 0-1-0, three Finnish 4-6, two Rumanian 3-2, one Rumanian 2-2.

(1.63) Reinforcements should be given in a fashion similar to the starting units. If Reinforcements are to appear in a specific hex and on a specific Game-Turn, use the following arrangement:
EXAMPLE:
One Game-Turn Two (in 2203)
One 4-4, two 3-4, six 8-6.

On Game-Turn Five (in 2201)
one 4-4.

Use the following style if several groups appear on the same Game-Turn but in different hexes:

One Game-Turn Two:
In 2203: One 4-4, two 3-4.
In 2230: six 8-6
In 3030: one 0-1-0, twenty-one AP

(1.64) Short special rules that apply only to one scenario should be given as part of that Scenario under the heading of "Special Rules." Special Rules that apply to two or more scenarios should be broken out into a separate Special Rules Main Section occuring after the Scenario Section (make reference to such Special Rules, by number in the affected scenarios).

(1.65) Use the following case numbering structure in scenarios whenever feasible:
EXAMPLE:
(17.2) SCENARIO NUMBER TWO: The Goon War, 1871-3
(17.21) Starting Forces and Deployment by Hex Number: The

Goon Army: (Phases First)
Four 3-3 (0101); two 8-8 (0201); etc, etc.

The Calgonites:
Two 1-1 (2021); four 5-2 (2123); etc, etc.

(17.22) Reinforcements:
The Goon Army:
On Game-Turn Twelve:
In 0101: one 3-3.
In 0201: four 8-1, three 2-10.

The Calgonites:
None.

(17.23) Special Rule
A. First special rule.
B. Second special rule.
C. Third special rule.

(17.24) Victory Conditions: The Goons win if they are the last to occupy or pass through the hex containing Wonkle City South (2502). The Calgonites win if they prevent this and/or if they exit three or more 5-2 units off the north map edge on or before Game-Turn Twenty-nine.

(17.25) Player's Notes: Copy copy copy copy.
(17.26) Historical Notes: Copy copy copy copy.

(1.66) Note that if there are few scenarios, Player's Notes and Historical Notes can be broken out into a separate Main Section dealing with the entire system. Such a section would also incorporate Developer's Notes and Designer's Notes. Whatever the case, these notes should not be just swamp gas. They should be informative, concise, accurate to the final game-system and components, and devoid of high school humor and obvious, smart-guy remarks.

(1.67) Consult the Art Director for advice on other formats.